W9-AHA-570

Esri Press
REDLANDS | CALIFORNIA

>>>
# PYTHON®
## Scripting for ArcGIS®

Paul A. Zandbergen

Esri Press, 380 New York Street, Redlands, California 92373-8100

Copyright © 2013 Esri
All rights reserved. First edition
Printed in the United States of America

17 16 15 14 13     1 2 3 4 5 6 7 8 9 10

*Library of Congress Cataloging-in-Publication Data*

Zandbergen, Paul A., 1968-
   Python scripting for ArcGIS / Paul A. Zandbergen.—First edition.
           pages cm
       Includes index.
       ISBN 978-1-58948-282-1 (pbk.)
       1. ArcGIS. 2. Geographic information systems. 3. Graphical user interfaces (Computer systems) 4. Python (Computer program language) I. Title.
   G70.212.Z36 2013
   910.285'5133--dc23                                    2012020676

The information contained in this document is the exclusive property of Esri unless otherwise noted. This work is protected under United States copyright law and the copyright laws of the given countries of origin and applicable international laws, treaties, and/or conventions. No part of this work may be reproduced or transmitted in any form or by any means, electronic or mechanical, including photocopying or recording, or by any information storage or retrieval system, except as expressly permitted in writing by Esri. All requests should be sent to Attention: Contracts and Legal Services Manager, Esri, 380 New York Street, Redlands, California 92373-8100 USA.

The information contained in this document is subject to change without notice.

U.S. Government Restricted/Limited Rights: Any software, documentation, and/or data delivered hereunder is subject to the terms of the License Agreement. The commercial license rights in the License Agreement strictly govern Licensee's use, reproduction, or disclosure of the software, data, and documentation. In no event shall the US Government acquire greater than RESTRICTED/LIMITED RIGHTS. At a minimum, use, duplication, or disclosure by the US Government is subject to restrictions as set forth in FAR §52.227-14 Alternates I, II, and III (DEC 2007); FAR §52.227-19(b) (DEC 2007) and/or FAR §12.211/12.212 (Commercial Technical Data/Computer Software); and DFARS §252.227-7015 (DEC 2011) (Technical Data – Commercial Items) and/or DFARS §227.7202 (Commercial Computer Software and Commercial Computer Software Documentation), as applicable. Contractor/Manufacturer is Esri, 380 New York Street, Redlands, CA 92373-8100, USA.

@esri.com, 3D Analyst, ACORN, Address Coder, ADF, AML, ArcAtlas, ArcCAD, ArcCatalog, ArcCOGO, ArcData, ArcDoc, ArcEdit, ArcEditor, ArcEurope, ArcExplorer, ArcExpress, ArcGIS, arcgis.com, ArcGlobe, ArcGrid, ArcIMS, ARC/INFO, ArcInfo, ArcInfo Librarian, ArcLessons, ArcLocation, ArcLogistics, ArcMap, ArcNetwork, *ArcNews*, ArcObjects, ArcOpen, ArcPad, ArcPlot, ArcPress, ArcPy, ArcReader, ArcScan, ArcScene, ArcSchool, ArcScripts, ArcSDE, ArcSdl, ArcSketch, ArcStorm, ArcSurvey, ArcTIN, ArcToolbox, ArcTools, ArcUSA, *ArcUser<,ital>*, ArcView, ArcVoyager, *ArcWatch*, ArcWeb, ArcWorld, ArcXML, Atlas GIS, AtlasWare, Avenue, BAO, Business Analyst, Business Analyst Online, BusinessMAP, CityEngine, CommunityInfo, Database Integrator, DBI Kit, EDN, Esri, esri.com, Esri—Team GIS, Esri—*The GIS Company*, Esri—The GIS People, Esri—The GIS Software Leader, FormEdit, GeoCollector, Geographic Design System, Geography Matters, Geography Network, geographynetwork.com, Geoloqi, Geotrigger, GIS by Esri, gis.com, GISData Server, GIS Day, gisday.com, GIS for Everyone, JTX, MapIt, Maplex, MapObjects, MapStudio, ModelBuilder, MOLE, MPS—Atlas, PLTS, Rent-a-Tech, SDE, SML, Sourcebook•America, SpatiaLABS, Spatial Database Engine, StreetMap, Tapestry, the ARC/INFO logo, the ArcGIS Explorer logo, the ArcGIS logo, the ArcPad logo, the Esri globe logo, the Esri Press logo, The Geographic Advantage, The Geographic Approach, the GIS Day logo, the MapIt logo, The World's Leading Desktop GIS, *Water Writes*, and Your Personal Geographic Information System are trademarks, service marks, or registered marks of Esri in the United States, the European Community, or certain other jurisdictions. CityEngine is a registered trademark of Procedural AG and is distributed under license by Esri. Other companies and products or services mentioned herein may be trademarks, service marks, or registered marks of their respective mark owners.

Ask for Esri Press titles at your local bookstore or order by calling 800-447-9778, or shop online at esri.com/esripress. Outside the United States, contact your local Esri distributor or shop online at eurospanbookstore.com/esri.

Esri Press titles are distributed to the trade by the following:

In North America:
Ingram Publisher Services
Toll-free telephone: 800-648-3104
Toll-free fax: 800-838-1149
E-mail: customerservice@ingrampublisherservices.com

In the United Kingdom, Europe, Middle East and Africa, Asia, and Australia:
Eurospan Group          Telephone: 44(0) 1767 604972
3 Henrietta Street      Fax: 44(0) 1767 601640
London WC2E 8LU         E-mail: eurospan@turpin-distribution.com
United Kingdom

# Contents

# Preface

The impetus for writing this book came from the lessons I've learned from using and teaching geographic information systems (GIS) for over 10 years at several different universities. One of these lessons is that "a little bit of code can go a long way."

Those of us who learned our first computer skills back in the days of MS-DOS became familiar with using a command prompt to carry out basic tasks. Early versions of ArcGIS for Desktop Advanced (ArcInfo) software also relied heavily on a command line interface, in addition to the use of the ARC Macro Language (AML). More recently, however, most software has come to rely on a graphic user interface (GUI) with very limited need to use a command line interface or to write any code. As a result, the majority of college students taking their first GIS course today have never seen any form of code. Although the menu-driven user interface of ArcGIS for Desktop allows for very complicated operations and sophisticated spatial analysis, at some point users will run into tasks that require something more. That's where Python scripting comes in.

In a nutshell, Python scripting allows you to automate tasks in ArcGIS that would be quite cumbersome using the regular menu-driven interface. For example, consider having to convert 1,000 shapefiles into feature classes in a geodatabase. You could run the appropriate tool 1,000 times, but surely there must be a more efficient and robust way to do this. That's what Python scripting will do, and you need only a handful of lines of code to carry out this task. About halfway through this book, you will write a script that does exactly that.

This book is designed to make the power of Python scripting available to those who have no experience writing code. The book starts with the basics, such as what scripting is and how to write and run simple lines of code. Following this, the book covers how to write scripts that work with spatial data in ArcGIS for Desktop applications. A good familiarity with ArcGIS for Desktop is assumed, including managing data in the ArcCatalog application and carrying out basic tasks in the ArcMap application, such as manipulating data, creating cartographic output, and running tools. You should also be familiar with the basic concepts of GIS, including coordinate systems, data formats, table operations, and basic spatial analysis methods. If you have some experience in writing code in any programming language, it will be helpful but is not required.

Why Python? For a couple of reasons. First, Python is free and open source, meaning it can be freely distributed and shared. Second, it is a powerful and versatile programming language although still relatively easy to learn. Third, Esri has adopted Python as the preferred language for working with ArcGIS, which is strongly reflected in the functionality introduced in version 10.

Although Python is included in a typical installation of ArcGIS for Desktop, it is important to remember that Python was not developed by Esri. There is a large and active Python community that maintains and continues to improve Python. As you learn the fundamentals of Python in this book, the additional benefit is that you may find other uses for Python beyond working with ArcGIS. Many introductory computer science courses now use Python as a way to explain fundamental concepts in computer programming. This book will introduce you to some of these fundamentals, but the emphasis is clearly on writing code that is specifically designed to work with spatial data in ArcGIS.

There are numerous introductory textbooks on GIS and tutorials for learning ArcGIS. Most of them include sections on spatial analysis methods and procedures. However, coverage of Python scripts is not at all complete. Although there is no lack of good introductory books on Python, most of them cover Python without targeting a specific application. Python's role as a "glue" language is explained and demonstrated, but most books remain very general when it comes to how Python works with other programs.

There is no ArcGIS-specific version of the Python language, so you can use any of the general Python books to start learning Python syntax. However, the objects you work with in ArcGIS, such as feature classes, polygons, and geoprocessing tools, are very different from the more generic objects used as examples in most existing titles. This makes it difficult for an experienced ArcGIS user to just pick up a general Python book and start writing scripts for ArcGIS. For an experienced programmer who has previously programmed with ArcGIS in either VBA or C++, a generic Python reference might suffice, but other professionals will benefit greatly from a book that covers Python with a very specific focus on writing scripts for ArcGIS.

The primary audience for this book consists of experienced ArcGIS users who want to learn Python but have limited programming or scripting experience. Prior experience with other scripting or programming languages, such as Perl, VBA, VBScript, Java, or C++, is helpful but not required. More experienced programmers will also benefit, but the emphasis is on making Python scripting available to the large number of ArcGIS for Desktop users who want to get more out of the functions of ArcGIS without having to become full-time programmers and learn C++. Readers will be expected, however, to have good overall ArcGIS skills and a basic understanding of geoprocessing procedures.

This book is also intended for upper-division undergraduate and graduate courses in GIS. A handful of colleges and universities teach upper-division courses in GIS programming and in scripting, and this number is expected to increase.

This book contains four major parts. Part 1 covers the fundamentals of geoprocessing in ArcGIS for Desktop as well as the fundamentals of the Python language. Depending on your background and experience, you may already be familiar with some or all of this material. Part 2 covers how to write scripts that work with spatial data. This is really the core of the book and includes chapters on running tools in Python, describing data, and manipulating and creating data. Part 3 explores a number of more specialized tasks, such as map scripting, debugging and error handling, and creating Python functions and classes. Part 4 goes into how to create a tool out of your script and how to share it with others. By the end of the book, you will be able to create custom tools that use Python scripting to automate basic tasks in ArcGIS for Desktop.

Each of the 14 chapters in this book is accompanied by an exercise chapter that reinforces the concepts covered in the chapter. All 14 exercise chapters are included on the Data and Exercises DVD that comes with the book. You should first read the chapter, and then complete the accompanying exercise before moving on to the next chapter. Depending on your learning style and familiarity with coding, you can try out some of the code in the chapters as you read them, but you can also read the entire chapter first, and then start the exercise. You should complete the chapters and exercises in order because the concepts introduced in each new chapter build on the preceding ones. Most of the exercises include challenges at the end, which allow you to practice your skills. Solutions for these challenges are also included on the Data and Exercises DVD.

To do the exercises in this book, you need to have ArcGIS 10.1 for Desktop software installed on your computer, or else download a trial version of the software. See appendix C for instructions on how to download the software. You will need the code printed on the inside back cover of this book to access the download site.

This book will teach you the techniques needed to automate tasks in ArcGIS. Perhaps by the end of the book you will become a Python aficionado, or perhaps you will simply be able to save yourself hours of work by using one of the scripts from the book. Whatever the case, beyond the specific skills of writing Python scripts for ArcGIS, you will also learn the basic logic of writing code. This will be helpful beyond the specific task at hand. My hope is that the book will contribute to demystifying what "writing code" really is for those who may be a little intimidated by it. And that it will show that writing code is not difficult to learn. Coding as an approach to solving problems is not only a very powerful tool, but is also applicable to many endeavors—I sincerely hope this book will allow you to experience the versatility of Python coding.

*Paul A. Zandbergen*
*Albuquerque, NM USA*

# Acknowledgments

This book would not have been possible without the support of many individuals.

First, I would like to recognize the numerous students in my courses who have endured my teaching over the years. They say you learn something best by trying to teach it to others and that has certainly been the case for me in trying to teach Python to aspiring GIS professionals. I will be indebted to you always.

The contributions of the staff at Esri Press cannot be underestimated. Their ongoing feedback throughout the writing and editing of the manuscript has been invaluable.

A number of other Esri staff members have also left their mark on the book, in particular David Wynne, Robert Burke, Jeff Bigos, and Bruce Harold. Their insider perspectives have made the book more accurate and more complete. Several outside peer reviewers have also contributed with valuable feedback. I would like to thank Timothy Nyerges, Professor, Department of Geography, University of Washington; Robert J. Hijmans, Associate Professor, Department of Environmental Science and Policy, University of California, Davis; John Lowry, Senior Lecturer, School of Geography, Earth Science and Environment, University of the South Pacific; and Chris Garrard, Programmer/Analyst II, Remote Sensing/Geographic Information Systems Laboratory, Utah State University, Logan.

I would also like to thank my parents, who always encouraged me to seek a career path that would allow me to fulfill my curiosity about the world while at the same time trying to make it a better place.

Most importantly, this book would not have been possible without the continued support of my family. Marcia, Daniel, and Sofia, thank you for believing in me and for allowing me to pursue my passions.

*Paul A. Zandbergen*
*Albuquerque, NM USA*

```
rt arcpy
port random
rom arcpy import env
env.overwriteoutput = T
inputfc = arcpy.GetPara
outputfc = arcpy.GetPar
utcount = int(arcpy.Ge
esc = arcpy.Describe(i
ist = []
mlist = []
= desc.OIDField
cpy.SearchCurs
next()

etValue(
d(id)
t()
)
```

# Part 1
## Learning the fundamentals of Python and geoprocessing

# Chapter 1
## Introducing Python

## 1.1 Introduction

Python is a programming language that is both simple and powerful. For those who have struggled with learning programming languages in the past, this may come as a pleasant surprise.

This chapter describes some of the main features of Python and its use as a programming language to write scripts for ArcGIS. The logic and structure of this book and the accompanying exercises is described, followed by some examples of how Python is used. The final part of this chapter introduces Python editors that make it easier to write and organize your code.

## 1.2 Exploring the features of Python

Python has a number of features that make it the programming language of choice for working with ArcGIS. Among them:

*It's simple and easy to learn:* Python is easy to learn compared with other highly structured programming languages like C++ or Visual Basic. The syntax is simple, which gives you more time to focus on solving problems than having to learn the language itself.

*It's free and open source:* Python is free and open source software (FOSS). You can freely distribute copies of the software, read the source code, make changes to it, and use pieces of it in new free programs. One of the reasons Python works so well is that it has been created, and is constantly being improved, by an active and dedicated user community. The FOSS nature of Python makes it possible for Esri to distribute Python with ArcGIS software.

*It's cross platform:* Python is supported on different platforms, including Windows, Mac, Linux, and many others. Python programs can work on any of these platforms with minimal change or sometimes no change at all. Since the ArcGIS for Desktop application runs only on Windows, this may not seem like a big advantage, but the user community for Python is large, in part due to its cross-platform nature.

*It's interpreted:* Many programming languages require that a program be converted from the source language, such as C++ or Visual Basic, into binary code that the computer can understand. This requires a compiler with various options. Python is an interpreted language, which means it does not need compilation to binary code before it can be run. You simply run the program directly from the source code, which makes Python easier to work with and much more portable than other programming languages.

*It's object oriented:* Python is an object-oriented programming language. An object-oriented program involves a collection of interacting objects, as opposed to the conventional list of tasks. Many modern programming languages support object-oriented programming. ArcGIS is designed to work with object-oriented languages, and Python qualifies in this respect.

# 1.3 **Comparing scripting vs. programming**

Although Python is a programming language, it is often referred to as a scripting language. So, what is the difference? In general, a *scripting* language refers to automating certain functionality within another program, while a programming language involves the development of more sophisticated multifunctional applications. Scripting is a programming task that allows you to connect diverse existing components to accomplish a new, related task. Scripting is the "glue" that allows you to put various existing elements together. Programming, on the other hand, allows you to build components from scratch, as well as the applications that incorporate these components. Languages that work with these low-level primitives and the raw resources of the computer are referred to as system languages. Examples of system languages include C++ and .NET languages. Scripting languages use built-in higher-level functions and mask the detail a system language deals with. Examples of scripting languages include Python, Perl, PHP, and Ruby.

Esri, for example, relies primarily on C++ as the programming language to create ArcGIS software and all the different components or *objects*, called ArcObjects, that you find in the software. You can then use C++ to write your own software that utilizes these same ArcObjects as well as create your own objects. However, you can also use scripting to access the existing functionality of ArcGIS and connect functions in new ways to extend that functionality.

One of the strengths of Python is that it is both a scripting and a programming language, although it does not have quite the depth of a system language like C++. You can use it for relatively simple scripts as well as more advanced programming tasks. The focus of this book is writing Python scripts to carry out tasks in ArcGIS. Python can also be used for application development, but these aspects of using Python are not addressed in this book. Python is used here as an interpreted language to work directly with the existing functions available in ArcGIS.

## 1.4 Using scripting in ArcGIS

ArcGIS 9 introduced scripting support for many popular scripting languages, including Python, VBScript, JavaScript, JScript, and Perl. ArcGIS is COM compliant, meaning that it uses the widely used Component Object Model (COM) software architecture. Scripting languages are able to access all the functions available in ArcGIS, including any extensions. This makes scripting a very attractive and efficient method for automating tasks. Although the same automation can be accomplished using a system language like C++ or a .NET language, scripting often requires much less effort.

Python scripting has become a fundamental tool for geographic information systems (GIS) professionals to extend the functionality of ArcGIS and automate workflows. Several years ago, probably the most widely used approach was to use the built-in VBA (Visual Basic for Applications) programming tools. Since then, however, Python has emerged as a robust complement and alternative to VBA programming. Starting with ArcGIS 10, the VBA development environment is no longer installed by default, and Esri is actively discouraging the continued use of VBA. Although application development will continue to employ languages such as C++ and .NET, Python has a number of advantages, especially for GIS professionals who are not full-time programmers.

Python is not the only scripting language that can be used with ArcGIS, but it has certainly become the most widely used. This is largely because Python has the ease of use of a scripting language, as well as the programming capability of a complete developer language. Python is included in a typical ArcGIS for Desktop installation. Python has also been directly embedded in many tools in ArcGIS for Desktop. The Spatial Statistics toolbox, for example, consists almost entirely of Python scripts—even though the casual user does not necessarily notice (or need to). ArcGIS 10 has seen further integration of Python within the ArcGIS interface, and Esri has officially embraced Python as the preferred scripting tool for working with ArcGIS. Additional enhancements have been introduced in ArcGIS 10.1.

## 1.5 Python history and versions

Python was created by Guido van Rossum at the Centrum voor Wiskunde en Informatica (CWI) in the Netherlands and first released in 1991. Van Rossum remains active in the ongoing development of Python but has been joined by a large number of contributors. Compared to other languages,

Python has gone through a limited number of versions, reflecting a philosophy of incremental change and backward compatibility.

Python's features include lists, dictionaries, and strings, as well as more advanced elements such as metaclasses, generators, and list comprehensions. The robustness of Python reflects the need to include the basic features all programmers need as well as the more desired advanced ones that are common in other, more complex programming languages.

The Python version that is recommended for use with ArcGIS 10.1 is Python 2.7. Although you can download and install any version of Python for free, the installation of ArcGIS 10.1 comes with Python 2.7.2. Although version 2.*x* still works fine and continues to be widely used, a major new version of Python was deemed necessary to remove a number of small problems that accumulated over the years and make the language even cleaner. Although there are a number of differences between versions 2.7 and 3.*x*, the basic structure of the language has not changed. Despite the availability of version 3.*x*, ArcGIS 10.1 currently works with version 2.7. If and when version 3.*x* is adopted as the preferred version for use with ArcGIS, various utilities are available to convert code between the two versions.

It should be noted that although the installation of ArcGIS comes with Python, Esri did not develop Python. Esri relies on the FOSS nature of Python to distribute it with ArcGIS as the recommended scripting language. In addition, Esri has created functionality in ArcGIS to make it easy to work with Python. However, Python is widely used for tasks other than writing scripts to work with ArcGIS. The added benefit is that as you start learning more about Python in this book, you will be able to start using it for both ArcGIS and other tasks as well.

## 1.6  About this book

*Python Scripting for ArcGIS* consists of two elements:

1.  The printed book, which covers the theory of using Python

2.  A set of exercises in digital format that accompany the printed book (on the disk in the back)

The printed book consists of 14 chapters that explain the structure and syntax of Python and illustrate how to write scripts for ArcGIS. Sample code is provided throughout the text, but the book itself does not provide detailed step-by-step instructions. The exercises on the accompanying DVD provide detailed instructions for hands-on learning. There is one exercise for each chapter in the book. You are encouraged to first read a chapter from the book, complete the corresponding exercise, and then move on to the next chapter. Most of the exercises include challenges at the end, which allow you to practice your skills. Solutions for these challenges are included on the Data and Exercises DVD.

## The structure of the book

Chapter 1 provides an introduction to Python and gives you the opportunity to get started using Python editors and the Python window in ArcGIS.

Chapter 2 introduces the ArcGIS geoprocessing environment, including the use of tools and the ModelBuilder application. Experienced ArcGIS users will be familiar with most of the material, but a good review will be beneficial. Knowing what is possible with the existing set of tools in the ArcToolbox window will be very helpful in writing effective scripts. Similarly, Python scripts and ModelBuilder are often used in combination, so a good knowledge of ModelBuilder is recommended to get the most out of Python scripting.

Chapter 3 describes the Python window in ArcGIS, which serves as an interactive interpreter. The Python window allows you to run one or several lines of Python code directly from within an ArcGIS for Desktop application. Code written in the Python window can be saved to a script, and existing code from a script can be loaded into the Python window.

Chapter 4 explains the fundamentals of the Python language, including the use of statements, expressions, functions, methods, modules, and controlling workflow, as well as best practices for writing scripts. Chapter 4 covers the basic syntax of Python for the novice Python user. Experienced Python users will be familiar with this material.

Chapter 5 describes the ArcPy site package that was introduced with ArcGIS 10. ArcPy includes numerous modules, classes, and functions, which provide an effective way to integrate ArcGIS and Python. Chapter 5 also includes working with data paths, environment settings, and licensing issues.

Chapter 6 includes techniques for data exploration, including describing data and data structures, as well as working with strings, lists, tuples, and dictionaries to characterize data prior to using it in other operations.

Chapter 7 introduces how to manipulate spatial data, including the use of cursors, searching for data, and working with tables and fields. The lessons in chapter 7 make it possible to use Python to run SQL queries on spatial data.

Chapter 8 describes how to work with the geometric properties of spatial objects, including how to utilize the properties of existing features and how to create new features.

Chapter 9 describes how to work with rasters in Python, including the ArcPy Spatial Analyst extension module. This module contains a number of specialized map algebra operators and other classes for using rasters for spatial analysis.

Chapter 10 describes the ArcPy mapping module for automating mapping tasks. This includes working with map documents, data frames, and layers, as well as exporting and printing maps.

Chapter 11 describes error-handling techniques for anticipating common errors and making your code more robust. The code debugging environment using the PythonWin Debugger, which allows you to test your code interactively, is also covered.

Chapter 12 demonstrates how to create custom functions and classes in Python. This makes it easier to organize more complex code and use parts of your code in multiple scripts.

Chapter 13 explains how to create custom script tools, which make Python scripts available as tools in the ArcToolbox interface. It is one of the preferred methods for sharing Python scripts with other users and also makes it easier to add a Python script as a tool to a larger sequence of operations in ModelBuilder.

Chapter 14 outlines strategies for sharing script tools with others, including how to organize your files, work with paths, and provide documentation for script tools.

## 1.7  Exploring how Python is used

This section uses several examples to illustrate how Python is used to create scripts. The examples were obtained from scripts created by Esri and the ArcGIS user community. One of the reasons for presenting these examples is for you to become more familiar with looking at Python code. One of the best ways to learn how to write code is to work with existing code. You are not expected to be able to understand the code at this point, but the examples will give you a flavor of what is to come.

### Example 1: Determining address errors

The AddressErrors script tool was created by Bruce Harold, an Esri employee. The AddressErrors tool inspects street centerlines for possible errors in address ranges associated with street segments. A polyline feature class is output that includes one feature for every error detected and contains attribution that profiles that error.

The script is made available as a tool in a toolbox. Although the script was written in Python, the functionality of the script can be accessed the same way as any other tool.

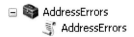

The tool dialog box looks like that of any regular geoprocessing tool.

The tool has 10 inputs, two of which are optional. The input features would typically consist of a polyline representing street centerlines and a number of attributes representing the range of street numbers. The output is a new feature class.

A single Python script is used in this tool, and the script can be opened to get an inside look at what the tool does. When you open the script in a Python editor, it looks like the one in the figure.

```
AddressErrors.py - C:\Scripts\AddressErrors.py

File  Edit  Format  Run  Options  Windows  Help

# Author: ESRI
# Date:     June 2010
#
# Purpose: This script checks street centreline data for errors in dual-range address attributes.
#          Errors reported are:
#
#              OVERLAP    - the address range overlaps the next segment
#              UNDERLAP   - the address range has a gap between the next segment
#              DIRECTION  - the segment range direction is opposite to the range origin
#              FROMTO     - the segment has a flipped from/to range
#              LEFTRIGHT  - the address ranges are on the wrong side
#              PARITY     - the address range disagrees with the assigned parity
#
#          Requires ArcGIS 10 - ArcInfo.
#
#

try:
    import arcpy
    import math
    import os
    import sys
    import traceback

    arcpy.env.overwriteOutput = True

    #Get the input feature class or layer
    inFeatures = arcpy.GetParameterAsText(0)
    inDesc = arcpy.Describe(inFeatures)
    if inDesc.dataType == "FeatureClass":
        inFeatures = arcpy.MakeFeatureLayer_management(inFeatures)
    searchRadius = str(inDesc.SpatialReference.XYTolerance * 10) + " " + \
                   str(inDesc.SpatialReference.LinearUnitName).replace('Foot_US','Feet')
    xyTol = inDesc.SpatialReference.XYTolerance
    inPath = os.path.dirname(inDesc.CatalogPath)
    sR = inDesc.spatialReference
    rangesAreText = False
```

```
Ln: 1  Col: 0
```

If you have not worked with Python or other programming languages before, this code can look a bit intimidating. However, the purpose of this book and the exercises that go with it is to make you familiar with Python syntax and the logic behind it for carrying out tasks in ArcGIS. So by the end of the book, you will be able not only to understand the preceding script, but also to write scripts of similar complexity.

## Example 2: Market analysis using the Huff Model tool

The Huff Model script was created by Drew Flater, an Esri employee. Here is a brief description of the tool as provided in the tool's instructions:

> *The Huff Model is a spatial interaction model that calculates gravity-based probabilities of consumers at each origin location patronizing each store in the store dataset. From these probabilities, sales potential can be calculated for each origin location based on disposable income, population, or other variables. The probability values at each origin location can optionally be used to generate probability surfaces and market areas for each store in the study area.*

This is a relatively sophisticated script, but like the first example, it is written entirely in Python. The script is also made available as a tool in a toolbox as shown in the figure.

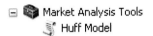

The tool dialog box has a lot of options for inputs, outputs, and analysis settings, reflecting the nature of the Huff Model.

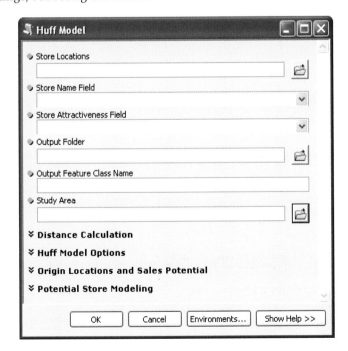

A tool Help file is provided that provides detailed descriptions of how the tool works, including the inputs, model formula, and outputs, as illustrated in the diagram that is part of the Help file and shown in the figure.

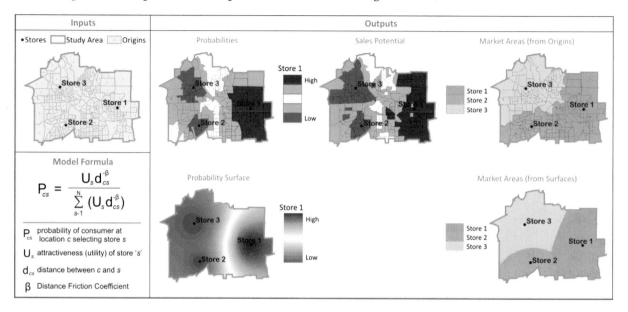

All the code for this tool is provided as a single Python script, part of which is shown in the figure.

```
HuffModel.py - C:\Scripts\HuffModel.py

File   Edit   Format   Run   Options   Windows   Help

# ------------------------------------------------------------------
# HuffModel.py
# Created: 4/13/2007 by Drew Flater
# Usage: Creating probability-based trade areas for retail stores
# ------------------------------------------------------------------

# Import system modules
import sys, string, arcgisscripting, os, traceback, shutil, re

# Create the Geoprocessor object
gp = arcgisscripting.create(93)

# Set overwrite
gp.overwriteoutput = 1

def AddPrintMessage(msg, severity):
    print msg
    if severity == 0: gp.AddMessage(msg)
    elif severity == 1: gp.AddWarning(msg)
    elif severity == 2: gp.AddError(msg)

# Start traceback Try-Except statement:
try:
    # Script parameters...
    stores = gp.getparameterastext(0)
    store_name = gp.getparameterastext(1)
    store_attr = gp.getparameterastext(2)
    outfolder = gp.getparameterastext(3)
    fc_name = gp.getparameterastext(4)
    studyarea = gp.getparameterastext(5)
    blockgroups = gp.getparameterastext(6)

Ln: 1  Col: 0
```

Given the complexity of the Huff Model, the Python script is quite lengthy, at over 700 lines of code. However, most of the tool's basic elements are the same as those found in simpler scripts. So as soon as you become familiar with what these elements are, you can start actually reading and understanding some of the more complex scripts.

Once you learn the basics, from the chapters in this book, of how to use Python for writing scripts, you will find one of the best ways to keep learning Python scripting is to work with existing code. Using example code can also speed up the process of writing your own scripts.

*Note: The two examples provided here were obtained from the Geoprocessing section of the ArcGIS Resource Center (http:// resources.ArcGIS.com). Of specific interest is the Geoprocessing Model and Script Tool Gallery (http:// resources.ArcGIS.com/ gallery/file/geoprocessing).*

# 1.8 Choosing a Python script editor

A computer script is essentially a list of commands that can be run by a certain program or scripting engine. Scripts are usually just plain text documents that have a specific file extension and contain instructions in a particular scripting language. Most scripts can be opened and edited using a basic text editor. However, using a specialized script editor to open a script provides additional functionality when you're writing the script and also allows you to run the script.

You can work with Python in a number of ways. The most basic approach is to use the so-called command line. If you have used other programming languages, you may be familiar with this type of interface. To access the Python command line in Windows, click the Start button, and then, on the Start menu, click All Programs > ArcGIS > Python 2.7 > Python (command line).

```
Python (command line)                                    _ □ ×
Python 2.7 (r27:82525, Jul  4 2010, 09:01:59) [MSC v.1500 32 bit (Intel)] on win
32
Type "help", "copyright", "credits" or "license" for more information.
>>> _
```

Although this interface provides full access to all of Python's functionality, it provides limited support for writing and testing scripts. So instead of using the command line interface, it is typically much more productive to use a Python script editor. A Python editor has a menu-driven interface and tools for organizing and testing Python scripts to make working with Python easier.

Python editors are also known as integrated development environments (IDEs). There are many different ones, including several open source and commercial packages. Some IDEs are developed for specific platforms, such as Windows, Mac, and Linux, and others are specifically designed to interface with particular programming languages, such as C++ and .NET languages. An extensive list of Python editors can be found on the Python wiki page (http://wiki.python.org/moin/PythonEditors). To a large degree, the

editor you use is a matter of preference, and experienced Python programmers all have their favorite ones. Python syntax remains the same for
different editors, which is one of the strengths of Python.

The default IDE that comes with any installation of Python is the
integrated development environment (IDLE). To access Python IDLE in
Windows, click the Start button, and then, on the Start menu, click All
Programs > ArcGIS > Python 2.7 > IDLE (Python GUI). GUI stands for
graphic user interface. IDLE is also known as the Python shell.

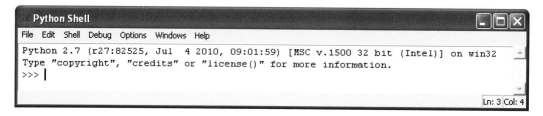

Descriptions of the various menu items can be obtained from the Help
menu. On the Python Shell menu bar, click Help > IDLE Help to view
these descriptions. More information on IDLE can be found at http://www.
python.org/idle.

Notice that the last line in the Python Shell window starts with > > >.
This is the *prompt* of the interactive Python interpreter. It is where you can
type code and press ENTER, and the interactive interpreter will carry out
your command. Are you ready for your very first line of Python?

```
>>> print "Hello World"
```

When you press ENTER, the following output appears:

```
Hello World
>>>
```

So, what's happening here? When you press ENTER, the interactive Python
interpreter reads the input command, prints the string `Hello World` to the
next line, and gives you a new prompt on the following line, waiting for the
next input. The term *printing* here refers to writing text to the screen.

You have now seen why Python is called an *interpreted* programming
language. When you are finished typing your commands and press ENTER,
the commands are interpreted and immediately carried out.

What if you type something else that does not make sense to the interactive Python interpreter? For example:

```
>>> I like eggs for breakfast
```

You get immediate feedback that the syntax is invalid:

```
SyntaxError: invalid syntax
>>>
```

You will also notice something else about the interactive Python interpreter. As soon as you type some text, it is given a color based on the nature of the input. For example, as soon as you type the word `print`, the word turns orange. Similarly, the string `"Hello World"` turns green. This is a way for the interactive Python interpreter to show how the input is being interpreted—in this case, orange is for statements and green is for strings. This is called *syntax highlighting* and is a helpful means of error checking as you write code. Be aware, however, that syntax highlighting conventions vary between Python editors, so don't get too used to a particular color scheme.

When you are learning the basics of Python syntax, it is useful to work directly in the interactive Python interpreter. You get immediate results that way, and you can keep going with new lines of code without worrying about having to save your work. However, when you are ready to write slightly more complex, multiline code, it will be more beneficial to write it as a script you can save. So remember that code written to the interactive Python interpreter is not meant to be saved.

Now you can see how writing a script differs from writing code in the interactive Python interpreter. On the Python Shell menu bar, click File > New Window. This opens a new window called Untitled. It is a script window, and there is no prompt.

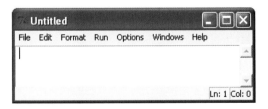

Now type the same line of code as before:

```
print "Hello World"
```

When you press ENTER, nothing happens—that's because a script itself is not interactive. A script needs to be *run* as a program for the command to be executed. Before you can run a script, however, it needs to be saved. On the menu bar, click File > Save As and save your script as **hello.py**. The file extension .py indicates it is a Python script. Now it is time to run the script. On the menu bar, click Run > Run Module. The string "Hello World" is printed to the interactive Python interpreter.

When you are coding, it is useful to have both the interactive Python interpreter and your script(s) open at the same time. If you want to try out something very quickly or check the syntax of a particular line of code, you can use the interactive Python interpreter. The script window contains the actual lines of code you want to save and keep working on. Occasionally, you can test your script by producing results that are printed to the interactive Python interpreter.

One widely used Python editor on the Windows platform is PythonWin. For the remainder of this book, it is assumed that you have access to PythonWin. The syntax, however, is the same, independent of the editor. The major differences between editors lie in how Python scripts are created, organized, and tested, although the syntax is the same.

> **Note:** *PythonWin is not installed by default when Python is installed during the ArcGIS installation. The corresponding exercise chapter, Exercise01, on the disk that comes with this book, includes instructions on how to install PythonWin.*

The basic PythonWin interface is shown in the figure. By default, it opens to an interactive Python interpreter called the Interactive Window, just like Python IDLE.

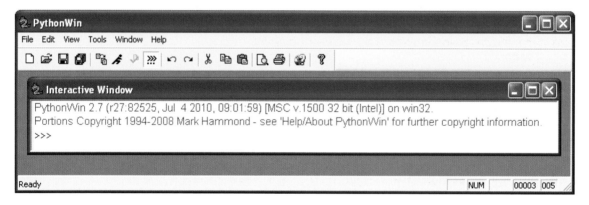

To create a new script, on the PythonWin menu bar, click File > New > Python Script. A new script window appears that can be resized so that both your interactive window and the script window are visible. To save the script, on the script window menu bar, click File > Save As. In the example that follows, the script is saved as `hellowin.py`. In the script window, enter the following line of code:

```
print "Hello World"
```

To run the script, on the script window menu bar, click File > Run. The result is printed to the Interactive Window. Notice that the syntax highlighting in PythonWin is slightly different.

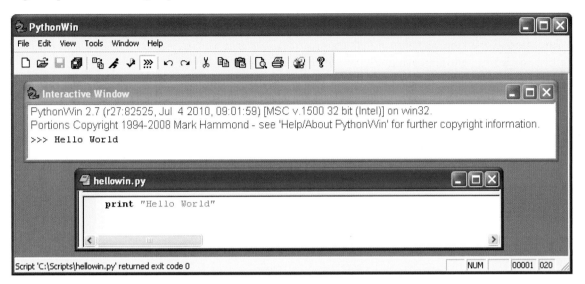

> **Note:** *From now on, the examples in the book use PythonWin as the Python editor. Choosing a Python editor is largely up to you. PythonWin was selected for this book because it provides a relatively easy-to-use yet robust and proven environment for working with Python on the Windows platform. Experienced coders can use the code examples in this book in their preferred code editor, as long as it is Python compatible.*

In addition to dedicated Python editors such as IDLE and PythonWin, general-purpose code editors can also be used for writing Python scripts. These editors usually offer syntax highlighting, formatting options, and other types of options for working with code. Examples of widely used code editors include Bluefish, Context, and Notepad++. Typically, these code editors work with multiple programming languages, and as a result, experienced code developers may use a single editor to work with multiple languages. In contrast, PythonWin is designed to work exclusively with Python code. Dedicated Python editors have some advantages over text editors, including code autocompletion prompts and debugging procedures, which is one of the reasons why the PythonWin editor is used in this book rather than a more generic code editor.

> **Note:** *Be careful using basic text editors, such as Notepad, that are not designed specifically for writing code. These editors do not preserve the proper formatting for a script and do not provide syntax highlighting or any other tools that help to write properly formatted scripts. Make sure a text editor is specifically designed for working with code such as the examples listed here.*

Finally, starting with ArcGIS 10, there is a way to use Python directly from within ArcGIS for Desktop applications that is both convenient and effective. The Command Line window in ArcGIS 9 has been replaced in ArcGIS 10 by the new Python window. When you open the Python window, you have an interactive interpreter for Python on the left and a Help section on the right.

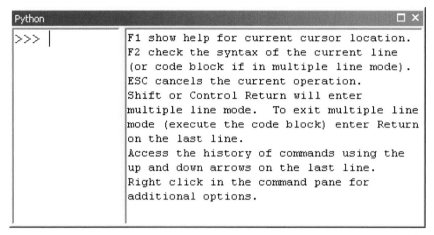

You can enter code in the Python window and each line is executed immediately, as in any other interactive interpreter. →

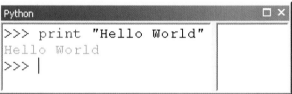

The Python window allows for rapid testing of simple lines of code. One advantage is that it provides syntax suggestions, known as code auto-completion "prompts," as you type. For example, when you type the letter *p*, the Python window suggests the statements `pass` and `print` to complete the code. This allows for faster coding and fewer typos when the code is completed for you. →

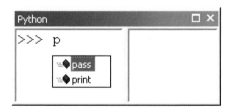

The most effective use of the Python window is covered in greater detail in later chapters.

To review, Python scripts, by definition, are modules, which is Python's highest level of organizing code. The actual file that contains the Python script is a simple text file, and therefore any text editor can be used to write a script. The file extension for a Python script is .py, which is automatically associated with a Python editor.

> **Note:** *Python script names must start with a letter and can be followed by any number of letters, digits, or underscores (_). Script names should not use Python keywords—you will see later what those keywords are.*

***Convention on sample code:*** Whenever sample code in the book is preceded by the prompt ( > > > ), the code is being written to an interactive interpreter (for example, the Python window in ArcGIS for Desktop applications or the Interactive Window in PythonWin). When you press ENTER, the line of code is executed immediately. Whenever sample code is not preceded by the prompt, the code is being written in a script window and the script needs to be run for the code to be executed. Much of the sample code, however, can be run in both an interactive interpreter and a script.

*Note: This book uses the PythonWin editor and the Python window in ArcGIS for Desktop applications to write and run code in Python. However, all the code can be used in any editor.*

## Points to remember

- ArcGIS for Desktop supports the use of scripting languages to automate workflows. Python is the preferred scripting language for working with ArcGIS.

- Python is not created by Esri. It is an open source programming language and therefore can be distributed by third parties, including Esri.

- Python is relatively easy to use. There is a large user community and there are many resources for learning Python. There is also a growing set of libraries for use in Python that provide additional functionality.

- One of the strengths of Python is that it is both a scripting language and a programming language. So you can use it for relatively simple scripts as well as for more advanced programming tasks. Using Python to develop scripts for ArcGIS is the focus of this book.

- Python is an interpreted language, which means it does not need to be compiled. Python scripts are run directly from the source code, making Python easier to work with and more portable than code in compiled languages such as C++ and .NET languages.

- Python scripts can be integrated into ArcGIS as script tools, which work just like the familiar geoprocessing tools.

- Working with Python requires the use of an editor, such as a general-purpose code editor or a specific Python editor. The default Python editor that installs with Python is called IDLE. This book employs PythonWin as the Python editor because it is relatively easy to use on the Windows platform. ArcGIS for Desktop applications also contain the Python window, which works like an interactive interpreter for Python.

- Python is installed as part of a typical ArcGIS for Desktop installation. This includes the IDLE editor by default, but not PythonWin.

# Chapter 2
## Geoprocessing in ArcGIS

## 2.1 Introduction

This chapter introduces the ArcGIS geoprocessing framework, including the use of ArcToolbox, ModelBuilder, and Python. Experienced ArcGIS users will be familiar with most of this material, but a review is beneficial. Understanding the geoprocessing framework is helpful in writing effective geoprocessing scripts. Similarly, Python and ModelBuilder are often used in tandem, so a good knowledge of ModelBuilder is recommended to get the most out of Python scripting.

## 2.2 What is geoprocessing?

Geoprocessing in ArcGIS allows you to perform spatial analysis and modeling as well as automate GIS tasks. A typical geoprocessing tool takes input data (a feature class, raster, or table), performs a geoprocessing task, and produces output data as a result. ArcGIS contains hundreds of geoprocessing tools. Examples of geoprocessing tools include tools for creating a buffer, for adding a field to a table, and for geocoding a table of addresses.

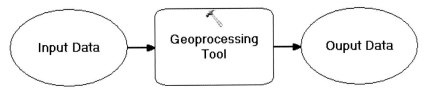

Geoprocessing supports the automation of workflows by creating a sequence that combines a series of tools. The output of one tool effectively becomes the input of the next tool. Creating these automated workflows combining geoprocessing tools can be accomplished in ArcGIS through the use of models and scripts.

The geoprocessing framework in ArcGIS consists of a set of windows and dialog boxes that are used to organize and execute tools. The geoprocessing framework makes it easy to create, execute, manage, document, and share geoprocessing workflows. Geoprocessing includes a set of tools that operate on data. The basic geoprocessing framework comprises the following:

- A collection of tools, organized in toolboxes and toolsets

- Methods to find and execute tools, including the Search window, the Catalog window, and the ArcToolbox window

- Tool dialog boxes for specifying tool parameters and executing tools

- ModelBuilder for creating models that allow for the sequencing of tools

- A Python window for executing tools using Python

- A Results window that logs the geoprocessing tools being executed

- Methods for creating Python scripts and using them as tools

Each of these components is described in more detail in the sections that follow. A few characteristics of this geoprocessing framework make it possible to work with tools in a consistent yet flexible manner. These characteristics include the following:

- All tools can be accessed from their toolbox, which makes for a consistent procedure for accessing tools, models, and scripts.

- All tools are documented the same way, which allows for consistent cataloging and searching.

- All tools have a similar user interface (the dialog box) for specifying the tool parameters.

- Tools can be shared.

# 2.3 A note on ArcObjects

You may recall the term "ArcObjects" from chapter 1. This section briefly outlines what ArcObjects are and how they relate to geoprocessing in ArcGIS.

The ArcObjects library consists of basic programming objects Esri has created to develop ArcGIS software. ArcObjects are made available to application developers as part of the ArcObjects .NET Software Development Kit (SDK) and the ArcObjects Java SDK. Application developers can use ArcObjects to create new applications or to enhance the functionality of existing ArcGIS applications. Esri software developers themselves use ArcObjects to create most of the tools in ArcGIS as well as to build the geoprocessing framework.

ArcObjects is intended for use with a system programming language—that is, a language that can access system-level functions to implement complex logic and algorithms. ArcObjects consists of thousands of different objects, which give a programmer a good degree of control over what the application is going to look like and how it is going to work. Examples of system programming languages include C++ and .NET languages, which are some of the most common languages for working with ArcObjects. These languages require substantial programming knowledge, much more than is required for working with models and scripts.

The ArcObjects SDKs and the geoprocessing framework are complementary, yet they accomplish different goals. ArcObjects is used to extend ArcGIS through new behaviors and to write stand-alone applications that build on the functionality of ArcGIS. Examples include creating new user interfaces and adding new behavior to feature classes. The geoprocessing framework is used to run existing tools and to create new tools (models and scripts) that automate tasks within the existing functionality of ArcGIS.

ArcGIS 10 introduced the desktop add-in model, which makes it possible to customize and extend ArcGIS for Desktop applications. In ArcGIS 10.1 for Desktop, Python was added to the list of languages that can be used to author desktop add-ins. Python add-ins can be used for some of the same things that were previously only possible using ArcObjects.

The emphasis in this book is using Python to create geoprocessing tools. ArcObjects is not used directly and is therefore not covered in this book. Desktop add-ins is also not covered. It should be noted that it is possible to work directly with ArcObjects using Python—it is a programming language, after all. However, the real strength of using Python lies in the ability to write powerful scripts, which requires a moderate level of programming skills and effort.

# 2.4  **Using toolboxes and tools**

Geoprocessing tools perform operations on datasets. Several hundred tools are available in ArcGIS. Exactly which tools you have available in ArcGIS depends on which product license you have (ArcGIS for Desktop Basic, ArcGIS for Desktop Standard, or ArcGIS for Desktop Advanced, formerly known as the ArcView license level, ArcEditor license level, and ArcInfo license level, respectively) and whether you have extensions installed (such as the ArcGIS 3D Analyst for Desktop extension, ArcGIS Network Analyst for Desktop extension, ArcGIS Spatial Analyst for Desktop extension, and others). The organization of the tools, however, remains the same. ➤

In ArcToolbox, geoprocessing tools are organized into toolboxes—for example Analysis Tools, Cartography Tools, and Conversion Tools, among others. Each toolbox typically contains one or more toolsets, and each toolset contains one or more tools. ➤

There are several ways to find the tools you need:

- Some of the most commonly used tools can be accessed directly from the Geoprocessing menu in ArcGIS for Desktop applications. Only a handful of tools are listed there.

- Another approach is to search for tools. In the Search window, you can type a term and search your map documents, data files, and tools. You can also filter your results to search for tools only. ➤

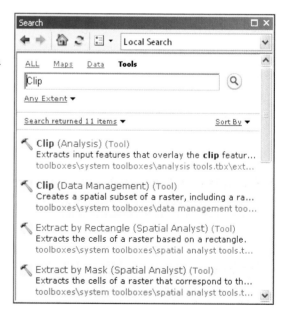

- To see all the available tools, you can open ArcToolbox and browse through the toolboxes and toolsets until you find the tool you want. Similarly, you can browse through the tools in the Catalog window in the ArcMap application or in the Catalog Tree window in the ArcCatalog application. Given the number of tools available, browsing can be cumbersome if you don't know where to look. Once you gain experience using the tools, however, you will start remembering where the tools are that you use most frequently. ➔

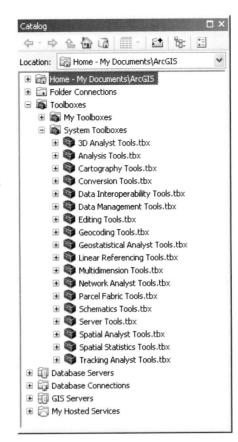

Once you have found a tool, you can double-click it to open the tool dialog box and fill in its parameters.

## 2.5 Learning types and categories of tools

There are four types of tools in ArcGIS, and each is designated by a different symbol.

*Built-in tools:* These tools are built using ArcObjects and a compiled programming language such as C++ or the .NET languages. Esri creates these tools when authoring its software, and most of the tools in ArcGIS look like this.

*Model tools:* These tools are created using ModelBuilder. A number of tools in ArcGIS are model tools—for example, some of the rendering tools in the Spatial Statistics toolbox.

*Script tools:* These tools consist of scripts that are accessible using a tool interface. When a tool is executed, a script is run to carry out the geoprocessing operations—for example, a Python script (.py), an AML (Arc Macro Language) file (.aml), or an executable file (.bat or .exe). Most of the script tools in ArcGIS use Python.

*Specialized tools:* These tools are created by system developers. They can have their own unique interface that is different from a regular tool dialog box. These tools are not very common, although third-party developers may distribute their tools in this manner.

Although these tools are created using different methods, the tool dialog boxes for different types of tools all look the same.

There are also two categories of tools:

1. *System tools:* These are the tools that are created by Esri and installed as part of the regular ArcGIS software. Exactly which system tools are installed depends on the product license level and the number of extensions. Most system tools are built-in tools, but a number of script and model tools are also authored by Esri.

2. *Custom tools:* These tools commonly consist of script and model tools, but built-in and specialized tools are also included. Custom tools can be created by a user, but they can also be obtained from a third party, and then added to ArcGIS.

When using geoprocessing tools, you may not notice which tools are system tools and which ones are custom tools because they are designed to work the same way. Once a custom tool is created, it can be added to a geoprocessing workflow the same way as any of the system tools.

## 2.6 Running tools using tool dialog boxes

When you find a tool, you can open it by double-clicking it, which brings up the tool dialog box. Each tool has a number of parameters that need to be specified before the tool can be run. A tool parameter is a text string, number, or other entry that tells the tool how it should be executed. The tool dialog box provides an easy-to-use interface for specifying these parameters. This includes browsing for and selecting datasets, selecting options from a list, and entering values.

Most tools require one or more input datasets. Other common parameters are preset text strings called keywords. Although each tool has one or more parameters, not all parameters are required. Optional parameters have default values that are set with the tool. You can accept the default values simply by not changing the parameters or by not specifying a value. Default values for keywords are typically shown on the tool dialog box when it is first opened.

An example of the tool dialog box for the Clip tool is shown in the figure.

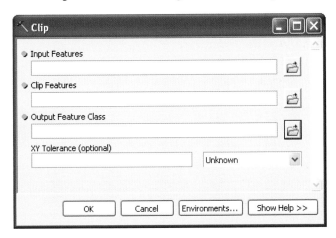

Every tool dialog box has a Help panel on the right side that provides useful information about the tool. You can switch the visibility of the Help panel by using the Show Help [ Show Help >> ] and Hide Help [ << Hide Help ] buttons at the bottom of the tool dialog box. To get a more complete description of the tool, you can click the Tool Help [ Tool Help ] button to access the tool's Help page. You will look at the Help pages in more detail in later chapters for examples of how to use a tool in a Python script.

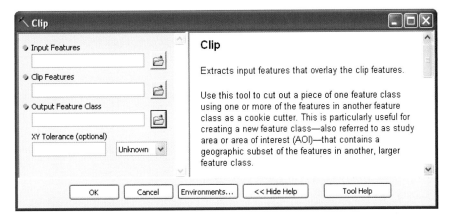

The tool dialog box contains the parameters that need to be specified for the tool to run. Notice that there are a total of four parameters in the case of the Clip tool. Three of them are flagged by a small green dot, which indicates the parameter is required and needs a value, and thus there is no default value. These are the Input Features to be clipped, the Clip Features, and the Output Feature Class to store the result. The XY Tolerance parameter is optional.

The tool dialog box has several mechanisms for ensuring proper inputs. For example, you could type the path and file name for Input Features (for example, C:\Data\streams.shp), but you could just as easily end up with a typo. Instead of typing a path and file name, you can click the drop-down arrow ▼ to select from a list of the layers in the table of contents in your current ArcMap document. This arrow is shown only when there are acceptable feature layers in your map document to choose from. You can also use the Browse button ⬚ to browse to data on disk. These options not only prevent typos, but also check for valid data input. For example, for the Clip tool, the Clip Features parameter has to consist of a polygon feature class, so the selection and browsing options will show only the available polygon feature classes.

Another feature of the tool dialog box is that the contents of the Help panel change depending on where your pointer is. When you first open the tool, a description of the tool appears in the Help panel. If you are not familiar with the tool, this description is useful for ensuring you have the correct tool. ➤

Once you click inside the input area for a particular parameter on the tool dialog box, the content of the Help panel changes to show an explanation of the parameter. For example, when you click in the XY Tolerance (optional) input area, a brief description is provided in the Help panel.

### Clip

Extracts input features that overlay the clip features.

Use this tool to cut out a piece of one feature class using one or more of the features in another feature class as a cookie cutter. This is particularly useful for creating a new feature class—also referred to as study area or area of interest (AOI)—that contains a geographic subset of the features in another, larger feature class.

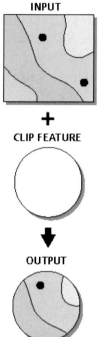

### XY Tolerance (optional)

The minimum distance separating all feature coordinates as well as the distance a coordinate can move in X or Y (or both). Set the value to be higher for data with less coordinate accuracy and lower for data with extremely high accuracy.

To get back to the overview Help, click anywhere on the tool dialog box, but not inside any parameter input areas.

Now consider an example dialog box that has the parameters completed. The input feature class is a shapefile (.shp) called roads.shp and is being clipped by a shapefile called zipcodes.shp. The output feature class is a shapefile called roads_clip.shp. The XY tolerance is left blank, which means the default value is used (which is 0.001 meters or its equivalent in map units for this parameter).

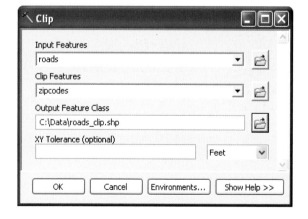

*Note: The full path to the output features is shown because the feature class was selected from disk. When selecting layers in the current ArcMap document using a drop-down list, only the name of a feature layer is shown and not its full path. In the latter case, the file extension, such as .shp, is not shown because the parameter is specified as a feature layer, not as a feature class on disk.*

When you click OK, the Clip tool runs. At the bottom of the ArcMap interface, a status bar displays the name of the tool that is being executed. By default, tool execution occurs in the background. This means you can continue working in ArcMap as the tool runs.

When the tool is finished running, a pop-up notification appears in the notification area, at the far-right corner of the taskbar. ➔

When a tool is finished running, the resulting feature class is added by default as a layer to the ArcMap table of contents (when the tool is run from within ArcMap). An entry is also posted to the Results window (on the menu bar, click Geoprocessing > Results). This entry includes all the input and output parameters, as well as tool execution messages. ➔

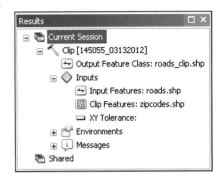

The entries in the Results window can be valuable in a number of ways. First, you can review the parameters that were used to run a particular tool. Second, you can run the same tool again directly from the Results window. The tool dialog box will be populated with the same parameters as before, and you can use these same parameters or change selected ones. Finally, you can review any error messages in the Results window.

Take a moment to review the tool parameters on a tool dialog box. Parameters that are required and need a value on the tool dialog box have a small green dot next to them. ➔

You can click the green dot to see more detailed information about the required parameter.

Optional parameters have no icon in front of them, and if they are left blank, the default values will be used when the tool runs. ➔

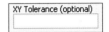

If an incorrect parameter is specified, an error warning appears. ➔

Pausing your pointer over the icon shows a brief description, while clicking the icon brings up a more detailed error message. In this case, the input dataset does not exist. ➔

ERROR 000732
Input Features: Dataset
C:\Data\streets.shp does not exist or is not supported

Sometimes a warning message appears, indicating that running the tool may lead to undesired results. ➔

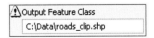

A warning message does not prevent the tool from running, but it may warrant taking a look at the warning prior to running the tool. In this case, the output dataset already exists and will be overwritten if the tool is executed. ➔

WARNING 000725
Output Feature Class: Dataset
C:\Data\roads_clip.shp already exists.

*Note: Overwriting geoprocessing results is an option under Geoprocessing > Geoprocessing Options. The default is turned off, meaning an attempt to overwrite existing datasets will result in an error. When the option is turned on, only a warning message is provided and the tool will run, overwriting the existing datasets.*

# 2.7 Specifying environment settings

Geoprocessing operations are influenced by *envi-ronment settings*. These settings are like additional hidden parameters that affect how a tool is run. The Environment Settings dialog box (Geoprocessing > Environments) allows you to view and set the environments for geoprocessing. →

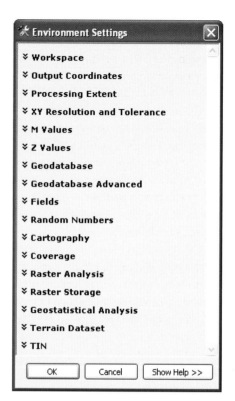

There are a number of settings, but one of the most important is the current workspace. Most geoprocessing tools use datasets as inputs, and then output new datasets. A workspace consists of a path to where these datasets are located. Complete path and file names can get quite long—for example, C:\Data\project_A12\water\final.gdb\roads\streets. To avoid having to type these lengthy names every single time (and possibly introduce typos), you can use the tool dialog box to select layers from the current ArcMap document or to browse to the location of a dataset. You can also drag files from the ArcMap table of contents to your map. In addition, a workspace can be set to make specifying input and output datasets easier. After your workspace is set, you need to specify only the base name. In the preceding example, you would set the workspace to C:\Data\project_A12\water\final.gdb\roads, and then enter only the base name "streets" when specifying tool parameters.

For example, how you would set the current workspace is shown in the figure. →

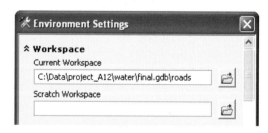

On a tool dialog box then, you could specify a feature class inside this workspace by typing only its base name. →

When you click anywhere else on the tool dialog box, the parameter is automatically completed using the current workspace, as shown in the figure. →

Output datasets are also created by default in the current workspace.

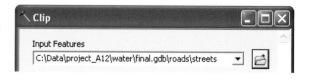

There are two types of workspaces: (1) the current workspace, which specifies, by default, where inputs are taken from and where outputs are placed; and (2) the scratch workspace, which is primarily used by model tools to write intermediate data.

There are also settings for specific data types (such as a geodatabase, raster, or TIN (triangulated irregular network)) and for specific types of functions (such as random numbers). Typically, you need to set only a few of these environments for a particular workflow because many of them do not apply to the data and tools you are using.

Environments are always at work. In other words, even if you don't specify them, they have default values that are used when a tool is run. For example, the default Output Coordinate System is the same one as the input. So when you are running a tool, the coordinate system is not changed, unless you specify otherwise on the Environment Settings dialog box.

Environment settings can be specified at a number of different levels, and there is a specific hierarchy to this process:

1.  The first level is the application. You can right-click in the ArcToolbox window and click Environments. This brings up the Environment Settings dialog box. Any settings created here are passed to the tools that are called by the application.

2.  The second level is the individual tool. Every tool dialog box has an Environments button Environments... . When you click the button, the Environment Settings dialog box opens. Any settings created here are applied only to the current running of the tool, and these settings override the settings passed by the application. These settings are not saved to the tool but apply only to a single execution of the tool.

3.  The third level is a model. Environment settings can be created as part of the model properties, which is separate from the settings you create on the tool dialog box. Any settings created in the model override the settings passed by the application or the tool dialog box settings. Model environment settings are saved as part of the model properties.

4.  The fourth and final level is a script. Environment settings can be coded into a Python script and these settings override the settings passed by the application or the tool dialog box. These settings are saved as part of the Python code in the script file.

In general, environment settings are passed down in this hierarchical system, but you can override these settings at each level.

# 2.8 **Using batch processing**

Most tools use a limited number of input datasets. For example, the Clip tool uses only a single input feature class and clips it using a single clip feature class. What if you wanted to run the same tool using similar settings on many different input datasets? This is where batch processing comes in. In the context of geoprocessing in ArcGIS, batch processing means executing a single tool multiple times using different inputs without further intervention.

All geoprocessing tools can be run in batch mode. Right-clicking a tool and clicking Batch brings up the Batch window of the particular tool. The Batch window shows a grid of rows and columns. The columns are the parameters of the tool and each row represents one execution of the tool. Rows can be added and the tool parameters can be specified for each run.

For example, in the case of the Clip tool, the Batch window looks like the example in the figure.

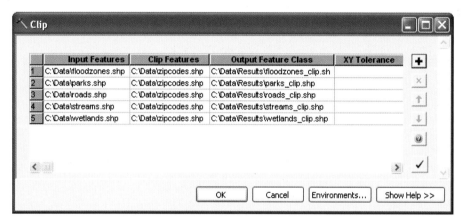

The batch grid contains five rows, indicating the Clip tool will run five times based on different inputs. The cells in each row represent the tool parameters. The buttons on the right allow for adding and deleting rows, changing the order of the rows, and examining the cells in the batch grid for valid values.

Entering parameters for the batch grid is similar to working with a regular tool dialog box, as follows:

- Click inside a cell and a drop-down arrow appears. This allows you to select from the layers in the table of contents of the current ArcMap document.

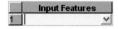

- You can drag layers from the table of contents into the dialog box.

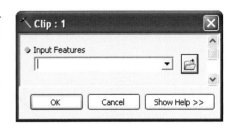

- Right-click in a cell and click Open. This brings up a separate dialog box that has the familiar drop-down arrow and Browse option. ➡

- Right-click in a cell and click Browse. This is a shortcut to the Browse option.

In addition to completing the batch grid cell by cell, there are a number of ways to quickly enter the parameter values for multiple cells, including (1) copying and pasting cell values and (2) using the Fill option to fill cells below a clicked cell with that cell's value.

One important feature of the Batch window is the Check Values button ✓. When you enter values into the batch grid, there is no automatic error checking. This is in contrast to regular tool dialog boxes, which automatically validate parameters—for example, checking whether a particular input dataset exists. When the Check Values button is clicked, all rows are scanned for errors and output dataset names are created if needed. If errors are found, the color of the cells changes.

Some common errors include the following:

- Green cells mean that a required parameter has not been specified.

- Red cells indicate an error was found and the tool will not run. The most common reason is that the input dataset does not exist.

- Yellow cells indicate a warning. The most common reason is that the output may not be what you expect.

Several other cell colors are also possible. White cells mean the parameters are correct. Gray cells mean the parameter is unavailable because it has no use, given the values of the other parameters. Finally, blue cells indicate a row is selected.

Running a tool in batch mode can be useful for setting up a large number of geoprocessing tasks. Although filling out the batch grid takes time, once the cell values are filled in, the tool runs in batch mode multiple times without additional user input. Running a tool in batch mode, however, does not reduce the time it takes for a tool to run. For example, running the Clip tool in batch mode using 20 rows takes the same amount of time as running the stand-alone Clip tool 20 times with the identical parameters. Time is saved by the quicker setup, not by faster tool execution.

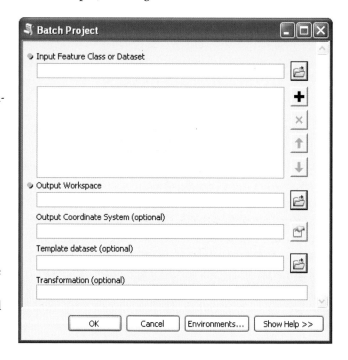

In addition to batch mode, there are a handful of specific batch system tools. For example, the Data Management toolbox contains the Project tool, which creates a new feature class with a different coordinate system from the input feature class. The Project tool uses only a single input feature class. The Batch Project tool, on the other hand, is the batch version of this tool and allows for multiple input feature classes. The same can be accomplished by running the Project tool in batch mode, but the parameter controls vary slightly. ➤

Both models and scripts provide additional ways to run batch processing, which is discussed in later chapters.

# 2.9 Using models and ModelBuilder

The execution of single tools is a practical way to accomplish certain GIS tasks. In a typical GIS workflow, however, you'll often need to run a sequence of tools to obtain the desired result. You could simply run through the sequence by running one tool at a time, but this has limitations, especially if your workflows are long and repetitive. ModelBuilder is one approach to creating this sequence of tools, whereby the output of one tool becomes the input to another tool. ModelBuilder is like a visual programming language—rather than using text-based instructions, it uses a visual flowchart to sequence geoprocessing tasks. A model in this context is a visual representation of a sequence of geoprocessing tasks. Within ArcGIS, models are tools, and once they are created, they work just like any other ArcGIS tool.

You can use any system or custom tool in your model, and there is no limit to how many tools you can use in a single model. Models can also include other models (since models are tools), and you can use iteration loops and conditions to control the flow of a model.

Before looking at how to create and run a model, first familiarize your-self with the basic elements of a model. Elements are the building blocks of a model. There are several types of elements: tools, data variables, value variables, and connectors.

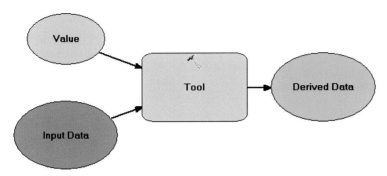

Geoprocessing *tools* are the basic building blocks of a model. Tools perform geoprocessing operations on geographic data. *Data variables* reference data on disk or a layer in the ArcMap table of contents. *Value variables* are items such as strings, numbers, Boolean values (True or False), spatial references, linear units, and extents. In short, value variables contain anything but refer-ences to data on disk. Variables are used as the input and output parameters of tools. Derived data, or the output variable of one tool, becomes the input variable of another tool. Data and values are connected to tools by *connec-tors*. The connector arrows show the direction of geoprocessing tasks. There are four types of connectors: (1) data connectors, which connect data and value variables to tools; (2) environment connectors, which connect a vari-able containing an environment setting to a tool; (3) precondition connectors, which connect a variable to a tool; and (4) feedback connectors, which con-nect the output of a tool back into the same tool as input. ➔

Connectors create model processes. A model process comprises a tool and the variables connected to the tool. The connector arrows specify the sequence of processing. A typical model contains a number of processes connected together. Complex models can contain hundreds of processes.

Creating a model and running tools in ModelBuilder consists of a num-ber of steps:

1. Create a new model.

2. Add data and tools to the model.

3. Create connectors and fill tool parameters.

4. Save the model.

5. Run the model.

6. Examine the model results.

*Create a new model:* There are two main ways to create a new model: (1) on the Standard toolbar in an ArcGIS for Desktop application, you can click the ModelBuilder button ; or (2) you can right-click a toolbox or toolset in ArcToolbox, and then click New > Model. This creates a new blank model in ModelBuilder.

*Add data and tools to the model:* You can add data and tools to the model either by dragging them from the ArcMap table of contents or ArcToolbox into the Model window or by using the Add Data or Tool button ✚ on the Model toolbar. In the model in the figure, a feature class called "roads" has been added as a data element and the Buffer tool has been added as a tool. Because the output of the Buffer tool is a new feature class, it has been automatically added as a derived-data element.

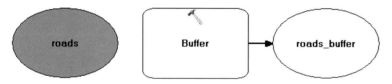

*Create connectors and fill tool parameters:* When you initially drag tools and data to a model, the process is not ready to run yet, because the required parameter values have not been specified. When any part of a process appears white in the model, it means that parameters are still missing. Parameters can be specified by opening the tool dialog box for each tool, and then specifying tool parameters as you would for any tool. Setting parameters creates connecting arrows, or connectors, between datasets and tools. You can also create these connectors using the Connect button ,which lets you select the specific parameter to be set for a particular tool. Once all the required parameters for a process are set, all the model process elements are turned into a specific color to show they're ready to run.

*Save the model:* You can save your model by clicking the Save button 🖫 or by clicking Model > Save on the Model toolbar. Model properties, including the name of the model and its display name, can be set by clicking Model > Model Properties.

*Run the model:* Once all the parameters are specified, the model is ready to run. You can run the entire model by clicking the Run button ▶ or by clicking Model > Run on the Model toolbar. You can also run a specific process by itself by right-clicking a tool and clicking Run. A Model progress dialog box indicates the progress made in running the tools included in the model. When the model run is completed, the model elements (other than the data inputs) have drop shadows to indicate the tools have been run and the output datasets have been created.

*Examine the model results:* By default, output datasets created by a model are considered intermediate, which means they are saved on disk but not automatically added to the ArcMap table of contents. To examine the model result, you can right-click the model element that contains the output dataset and click Add To Display. This adds the dataset to the ArcMap table of contents so you can examine the result.

Additional tools and data can be added to the model using the same steps. In the example model in the figure, the road layer is buffered and then intersected with a layer of geological hazard areas. The intersect result is then clipped using a watershed layer.

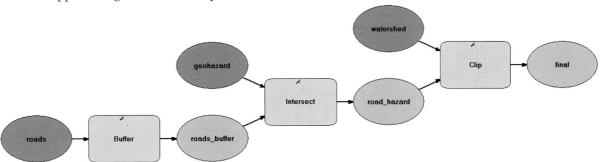

A model in ModelBuilder serves as a visual flowchart of a sequence of geoprocessing tools. The ModelBuilder interface provides an intuitive way to create this sequence. The key is that a model is a tool in a toolbox, which makes it possible to save the model for future use and to share it with others.

The model developed so far is relatively simple. However, there is no limit to the number of datasets and tools that can be used in a single model. Sophisticated models can contain a large number of geoprocessing tasks.

Among the advantages of using models to create geoprocessing workflows:

- ModelBuilder provides an intuitive interface to create workflows.

- Models provide an efficient mechanism to document workflows.

- Models can be organized in toolboxes and shared with others.

There are many more details to learn about ModelBuilder, which is beyond the scope of this book. ArcGIS Desktop Help provides extensive documentation on ModelBuilder. From within an ArcGIS for Desktop application, click Help > ArcGIS Desktop Help > Professional Library > Geoprocessing > Geoprocessing with ModelBuilder.

# 2.10  Using scripting

Just as ModelBuilder can be used to create models that run a sequence of tools, a scripting language can be used to create and run this sequence. Scripting languages are relatively easy to learn, and the primary scripting language used in ArcGIS is Python.

Scripts are analogous to models: ModelBuilder is used to create models, and Python is used to create scripts. ModelBuilder is a visual programming language, and Python is a text-based programming language. And just as models are tools within ArcGIS, scripts are also tools. So once a script is created, it becomes another tool you can run on its own or use in a model or in another script. Scripts can be run as stand-alone scripts on disk, in which case they are not a tool, but it is relatively easy to add a script as a script tool to a toolbox. Models can also be converted to scripts, but not vice versa. Converting a model to a script is covered in a later section of this chapter.

If models and scripts are so similar, why use a script instead of a model? ModelBuilder is a very intuitive way to create tools and relatively easy to learn for the beginning ArcGIS user. It requires no programming experience and there is no syntax to learn. Many geoprocessing workflows can be accomplished using models created in ModelBuilder. These models can be shared and modified. ModelBuilder, however, has certain limitations, and some of the more complex geoprocessing operations cannot be accomplished by a model alone. Some specific things you can do with a script that are not possible with a model include the following:

- Some lower-level geoprocessing tasks are possible only in scripts. For example, script cursors let you loop through records in a table, reading the existing rows and inserting new rows.

- Scripting allows for more advanced programming logic, such as advanced error handling and the use of more advanced data structures. Many scripting languages, including Python, have been extended with additional libraries offering more advanced functions.

- Scripting can be used to wrap other software—that is, to glue together applications. This facilitates the integration of various software applications. For example, Python can be used to access functions in Microsoft Excel or in the statistical package R.

- A script can be run as a stand-alone script on disk outside of ArcGIS. In most cases, you still need to have ArcGIS installed on the computer, but ArcMap or ArcCatalog do not need to be running for the script to work.

- Stand-alone scripts can be scheduled to run at a specific time without user intervention.

Python scripts can be created and run using a Python editor, such as PythonWin. You can also run Python code in the Python window of ArcGIS for Desktop. The Python window works like an interactive interpreter and code is executed immediately. The functionality of the Python window is discussed in more detail in chapter 3.

To run a tool in Python, type the tool name followed by its parameters. For example, the Python code shown in the figure runs the Clip tool.

```
Python
>>> import arcpy
>>> arcpy.Clip_analysis("C:/Data/roads.shp", "C:/Data/zipcodes.shp", "C:/Data/roads_clip.shp")
```

The result, which follows, is printed in the Python window and the resulting shapefile is added to the ArcMap table of contents.

```
<Result 'C:\\Data\\roads_clip.shp'>
```

Notice that the Python code in the figure (see preceding page) uses "arcpy" in the code. This is the ArcPy site package. The first line of code is `import arcpy`, which makes it possible to access all ArcGIS geoprocessing tools and other functionality in Python. ArcPy is described in detail in chapter 5. Don't worry too much about the exact syntax of the code for now. The basic idea is that you run a tool by typing the tool name followed by its parameters.

> **Note:** *ArcGIS 9 contained a Command Line window, which also allowed for the running of a tool by typing the tool name followed by the tool parameters. The text you entered was called a "command." The syntax of these commands, however, was specific to the ArcGIS environment—in fact, it relied heavily on the syntax of the old-style ArcInfo command line. You could not use Python in the Command Line window. In ArcGIS 10, the Command Line window has been replaced by the Python window. Those working with Python typically refer to "code" rather than "commands," but occasionally you will see references to Python "commands."*

Python code can be entered in the Python window and run immediately. You can also use a text editor or a Python editor to create and run Python files on disk. Python files have a .py extension and are known as scripts. Scripts are programs that can be run from the operating system, from a Python editor, or by using a script tool that runs the script. Following is an example of each.

The code contained in a script called clip_example.py is shown in the PythonWin editor. This is a modified version of the script that is provided within the Help page for the Clip tool. Again, don't worry too much about the exact syntax for now.

```
clip_example.py                                              _ □ ✕

    # Name: clip_example.py
    # Description: Script to execute a clip operation
    # Author: Esri

    # Import system modules
    import arcpy
    from arcpy import env

    # Set workspace
    env.workspace = "C:/Data"

    # Set local variables
    in_features = "roads.shp"
    clip_features = "zipcodes.shp"
    out_feature_class = "roads_clip.shp"
    xy_tolerance = ""

    # Execute clip
    arcpy.Clip_analysis(in_features, clip_features, out_feature_class, xy_tolerance)
```

You can navigate to the location of this script and double-click the file to run it. You do not need to have an ArcGIS for Desktop application open and you don't need to open the script in a Python editor. You can confirm the results of the script execution by examining the data in an ArcGIS for Desktop application. There are several benefits to running a script directly—most notably, you can set a script to run at a specific time without user intervention.

Another way to run a script is to use a Python editor like PythonWin. You can open a script in the editor, verify its content, and then execute the script. Similar to running a script directly from the operating system, you do not need to have an ArcGIS for Desktop application open for the script to run, although you need to have ArcGIS installed on your computer to be able to use the geoprocessing functions. One of the benefits of running a script using a Python editor is that messages are printed to the interactive window, including any error messages.

The third way to run a script is to create a script tool that runs the script. For example, you can create your own toolbox, create a new script tool (for example, My Clip Tool), and then add the clip_example.py script to this tool.

You can then run the script as you would any other geoprocessing tool. The benefit of running a script as a script tool from within ArcGIS is that you can integrate the script tool with other tools and models. The tool has its own dialog box, and the tool can be added to a model in ModelBuilder or called by another script.

The script used here so far is relatively simple and in fact does nothing more than the regular Clip tool. However, it is relatively easy to create scripts whose functionality exceeds that of existing tools—these scripts are covered in later chapters.

# 2.11 **Running scripts as tools**

As discussed in the previous section, scripts can be run in various ways. Running a script as a tool is a great way to integrate Python scripts into ArcGIS workflows. In fact, many scripts written by Esri are made available as tools in ArcToolbox. For example, take a look at the Proximity toolset within the Analysis toolbox. Notice that the Multiple Ring Buffer tool is a script tool, as shown in the figure. ➜

When you open the tool dialog box, it looks like a regular tool with several required and optional parameters. So from the perspective of a regular ArcGIS user, all tools in ArcToolbox look the same.

For most system tools in ArcGIS for Desktop, the underlying code cannot be viewed. However, for script tools, you can look "under the hood" by opening the script. To view the contents of a script, right-click the script tool and click Edit. This shows that the tool calls a script called MultiRingBuffer.py. These scripts are typically located in C:\Program Files\ArcGIS\Desktop10.1\ArcToolbox\Scripts. Several dozen of the system tools that come with ArcGIS for Desktop are script tools, and their content can be viewed in this manner.

The MultiRingbuffer.py script that is attached to the Multiple Ring Buffer tool is shown in the figure (see facing page).

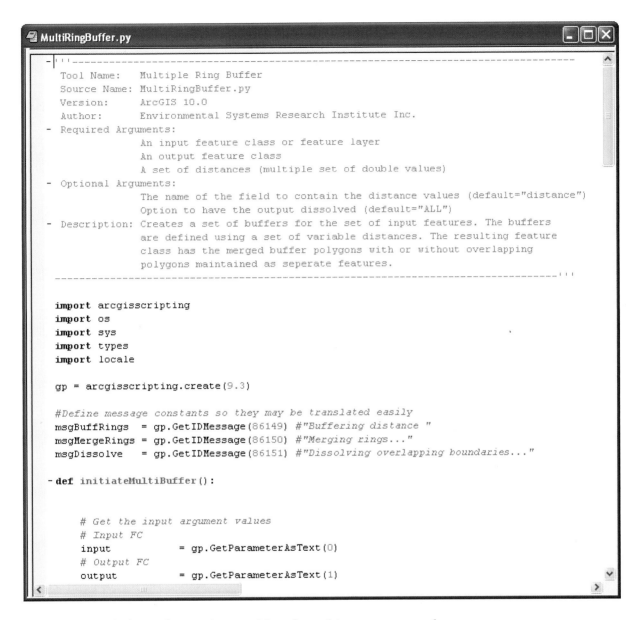

```
'''----------------------------------------------------------------------------
 Tool Name:    Multiple Ring Buffer
 Source Name:  MultiRingBuffer.py
 Version:      ArcGIS 10.0
 Author:       Environmental Systems Research Institute Inc.
 Required Arguments:
               An input feature class or feature layer
               An output feature class
               A set of distances (multiple set of double values)
 Optional Arguments:
               The name of the field to contain the distance values (default="distance")
               Option to have the output dissolved (default="ALL")
 Description: Creates a set of buffers for the set of input features. The buffers
              are defined using a set of variable distances. The resulting feature
              class has the merged buffer polygons with or without overlapping
              polygons maintained as seperate features.
------------------------------------------------------------------------------'''


import arcgisscripting
import os
import sys
import types
import locale


gp = arcgisscripting.create(9.3)

#Define message constants so they may be translated easily
msgBuffRings  = gp.GetIDMessage(86149) #"Buffering distance "
msgMergeRings = gp.GetIDMessage(86150) #"Merging rings..."
msgDissolve   = gp.GetIDMessage(86151) #"Dissolving overlapping boundaries..."

def initiateMultiBuffer():


    # Get the input argument values
    # Input FC
    input              = gp.GetParameterAsText(0)
    # Output FC
    output             = gp.GetParameterAsText(1)
```

Reading through this code can give you ideas for writing your own code.
The script is too lengthy to discuss in detail here. However, the basic idea
is that you can create a script and add it as a tool to a toolbox so that it
becomes a script tool a user can use without having to work directly with
the Python code.

Keep in mind that not all system tools are necessarily updated to
reflect the latest possibilities in Python scripting. For example, the
Multiple Ring Buffer script still uses the ArcGISscripting module,
the predecessor to the ArcPy site package. That is why you see the line
gp = ArcGISscripting.create(9.3), followed by frequent references to this geoprocessing object. Esri does not necessarily update all its
code since the tool works fine using the older ArcGISscripting module.

Tools that are altered substantially from earlier versions, however, are more likely to be rewritten specifically using ArcPy.

The script tools that are part of the system tools are read-only and cannot be edited. However, you can copy parts of the script code, or you can copy the script file itself to a different location and make edits to the script.

## 2.12 Converting a model to a script

Models and scripts are analogous in that they are both used to create a sequence of geoprocessing tasks. Models can be converted to Python scripts. On the ModelBuilder menu bar, click Model > Export > To Python Script.

Take a look at the model created previously, as shown in the figure.

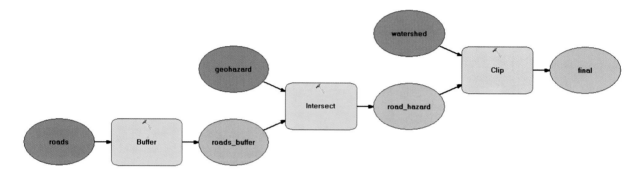

After exporting the model to a script, the Python code looks like the example in the figure.

```
# -*- coding: utf-8 -*-
# ------------------------------------------------------------
# model.py
# Created on: 2012-03-13 20:15:54.00000
#   (generated by ArcGIS/ModelBuilder)
# Description:
# ------------------------------------------------------------

# Import arcpy module
import arcpy

# Local variables:
roads = "C:\\Data\\roads.shp"
geohazard = "C:\\Data\\geohazard.shp"
watershed = "C:\\Data\\watershed.shp"
roads_buffer = "C:\\Data\\roads_buffer.shp"
road_hazard = "C:\\Data\\roads_hazard.shp"
final = "C:\\Data\\final.shp"

# Process: Buffer
arcpy.Buffer_analysis(roads, roads_buffer, "1000 Feet", "FULL", "ROUND", "ALL", "")

# Process: Intersect
arcpy.Intersect_analysis("C:\\Data\\roads_buffer.shp #;C:\\Data\\geohazard.shp #", road_hazard, "ALL", "", "INPUT")

# Process: Clip
arcpy.Clip_analysis(road_hazard, watershed, final, "")
```

The script contains all the elements from the model: the data inputs (roads, geohazard, and watershed) and the tools (Buffer, Intersect, and Clip).

Although a model can be exported to a Python script, however, the reverse is not true. Python scripts are more versatile than ModelBuilder, so that a Python script cannot be exported to a model.

Creating a model and converting it to a script is a good way to be introduced to what scripts look like and to get familiar with Python syntax. When a model is converted to a script, the Python code very closely follows the elements in the model, without adding anything extra. For example, the resulting script does not contain any specific validation or error-checking procedures. In general, converting a model provides a starting point for writing a script, including very specific code blocks, but it rarely results in a finished script.

## 2.13 Scheduling a Python script to run at prescribed times

Stand-alone scripts can be set to run at prescribed times. This can be useful for such things as carrying out routine data maintenance tasks. The steps for accomplishing this depend on the operating system.

*Step 1: Access the scheduled tasks.*

- For Windows XP: On the taskbar, click the Start button, and then, on the Start menu, click Control Panel > Scheduled Tasks. If the Control Panel is in category view, select Performance > Maintenance > Scheduled Tasks.

- For Windows Vista: On the taskbar, click the Start button, and then, on the Start menu, click Settings > Control Panel > System and Maintenance. Then click Administrative Tools > Schedule Tasks.

- For Windows 7: On the taskbar, click the Start button, and then, on the Start menu, click Control Panel > Administrative Tools > Task Scheduler. If the Control Panel is in category view, click System and Security > Administrative Tools > Task Scheduler.

*Step 2. Double-click Add Scheduled Task (or Create Basic Task).*

*Step 3. Complete the options on the wizard.*

When asked to click the program you want Windows to run, click the Browse button to select the Python script.

Many Python scripts require arguments to run. These can be set as part of the scheduled task. On the last dialog box of the wizard, select the "Open advanced properties" check box:

☑ Open advanced properties for this task when I click Finish.

On the dialog box that opens, the script to be run is shown in the Run box.

For a script to run with arguments, the Run box needs to be changed to a string that contains the Python executable file, the script, and the arguments to be passed to the script. For example, this would look like the following code:

```
c:\python27\python.exe c:\data\testscript.py c:\data\streams.shp
```

These arguments are similar to the way parameters are passed to a script from a script tool. If all information needed to run the script is hard-coded in the script itself, no arguments are needed.

Scheduling a Python script to run at prescribed times appears relatively simple, but there are some potential obstacles. First, the computer needs to be turned on for a scheduled task to be executed. Second, scheduled tasks typically require administrative access, and login information needs to be provided when the task is set up. Finally, many Windows-based PCs are configured to be locked or to log off current users after a certain period of inactivity, which can interfere with running scheduled tasks. So before you can rely on scripts being run in this manner, it is worthwhile to test your computer configuration to ensure scheduled tasks are run properly.

# Points to remember

- The geoprocessing framework in ArcGIS provides a powerful yet flexible system for organizing and running tools.

- ArcGIS has a large number of tools, organized in toolboxes and toolsets within ArcToolbox. The different types of tools include built-in tools, script tools, model tools, and custom tools.

- A tool is run by specifying tool parameters, including input and output datasets, and other parameters that control how a tool is executed.

- Environment settings also control how tools are executed and can be set at different levels.

- You can create your own tools using models and scripts. Once you have created your own tools, they work exactly like regular tools.

- ModelBuilder provides a visual programming language for creating a sequence of geoprocessing tasks. Models act like a flowchart.

- Python provides a text-based programming language for creating a sequence of geoprocessing tasks. Python code can be run in the Python window directly within ArcGIS. Python scripts (.py files) typically consist of more complex code and can be executed in various ways: directly from within the operating system, using a Python editor like PythonWin, or from a script tool within ArcGIS.

- Both models and scripts work like tools within the ArcGIS geoprocessing framework. Models can be converted to a script, but scripts cannot be converted to a model. Converting a model to a script is a good way to get started writing scripts in Python.

# Chapter 3
## Using the Python window

## 3.1 Introduction

You have already seen a few examples of using the Python window as an interactive interpreter for Python code that can be used directly within any ArcGIS for Desktop application. This chapter describes how to use the Python window in a bit more detail.

## 3.2 Opening the Python window

The geoprocessing Command Line window from earlier ArcGIS releases has been replaced by the Python window. The Python window is a quick and convenient way to run geoprocessing tools while taking advantage of other Python modules and libraries. It also provides a great way for beginners to learn Python.

The Python window can be used to run one or more lines of Python code. It is a useful place to experiment with syntax and work with short lengths of code. Scripting ideas can be tested outside a larger script directly within an ArcGIS for Desktop application. The availability of Python functionality directly within ArcGIS for Desktop applications provides an efficient mechanism for accessing and executing geoprocessing tasks.

The Python window can be opened from any ArcGIS for Desktop application by clicking the Python window button 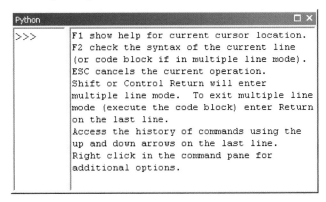 on the Standard toolbar. The Python window is shown in the figure. ➜

On the left side is the interactive Python interpreter, where Python code is entered. The primary prompt is indicated by three greater-than symbols (> > >). On the right side is the Help and syntax panel, which is updated as code is written.

The Python window can be docked or undocked and can also be resized. The vertical divider between the two sides can be moved. If you don't want the Help and syntax panel, the divider can be moved all the way to the right, as shown in the figure. To restore it, move the divider to the left. ➔

## 3.3 Writing and running code

Python code in the Python window is executed one line at a time, and the result is displayed immediately. Take a look at the example in the figure. ➔

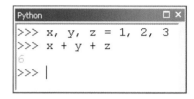

**Note:** *Python language fundamentals are covered in chapter 4. For now, don't worry about the exact syntax of Python code.*

When you press ENTER at the end of each line, the code is executed. Any result is printed to the next line and the following line starts with a new command prompt (> > >). In some cases, however, a single line of code cannot be executed because it is part of a multiline construct. Consider the example in the figure. ➔

The `if` statement is part of a multiline construct—there has to be at least one additional line of code for it to be executable. Therefore, the next line is a continuation line that uses the secondary prompt, which is three dots (...).

The `if` statement signifies the beginning of a block of code, and therefore the next line needs to be indented, like the example in the figure. ➔

Notice that the code has not been run, and the next line also starts with the secondary prompt. By default, the Python window does not assume the multiline construct is finished and you can continue writing the block of code without the code being executed.

When you are finished with the block of code and want to run it, simply press ENTER right after the secondary prompt. This runs the block of code, with any result printed to the next line, and a new primary prompt appears, like the example in the figure. ➔

The use of secondary prompts makes it possible to complete several lines of code before running it. You can force a secondary prompt by pressing CTRL+ENTER at the end of a line of code. After the first secondary prompt, you can press just the ENTER key to continue with the secondary prompt for the next line of code. To run the code, place the pointer right after the secondary prompt and press ENTER. ➔

An example of the use of a geoprocessing tool in the Python window is shown in the figure. The Get Count tool is used to determine the number of rows for a feature class or table. In this example, the feature class is obtained from a layer in the current map document and the result of the Get Count tool is printed.

```
Python                                                   □ ×
>>> import arcpy
>>> count = arcpy.GetCount_management("floodzone")
>>> print count
21308
>>>
```

Again, don't worry too much about the syntax at this point. The basic idea is that the Python window can be used to call any of the available geoprocessing tools. You can test code, practice Python syntax, and save your code, all from within ArcMap or ArcCatalog.

Initially, it may appear that running a tool from the Python window takes more effort compared with using the tool dialog box. However, there are many advantages to using the Python window:

- All geoprocessing tools can be accessed by importing the ArcPy site package. Other nontool functions such as listing and describing data, working with environment settings, and accessing geoprocessing messages are also available from the ArcPy site package.

- Autocompletion functionality can be a great assist in writing correct code, including the syntax needed for geoprocessing tools. Autocompletion is covered later in this chapter.

- Conditional execution can be accomplished using `if-then-else` logic.

- Iteration over many sets of data can be accomplished using `for` and `while` loops.

- Any functionality in standard Python modules can be accessed, including manipulation of strings, files, and folders.

- Python provides access to many third-party modules that expand the possibilities for manipulating data.

- Python blocks of code written in the Python window can be saved to a Python or text file for use in a Python editor, and code from existing script files can be loaded into the Python window.

# 3.4 **Getting assistance**

When working in the Python window, there are several shortcuts to get help with writing proper code:

- The F1 key: Pressing F1 shows the Help for the current pointer location. It is shown in the Help and syntax panel.

- The F2 key: Pressing F2 checks the syntax of the current line of code, or block of code for multiline constructs. Any errors are shown in the Help and syntax panel.

The UP ARROW and DOWN ARROW keys can be used to access previously entered commands. This is quite helpful when you have made a minor mistake—rather than retyping the entire command, you can simply bring up the previous line of code and make the change. Consider the example shown in the figure.

```
Python                                                              □ ×
>>> count = arcpy.GetCount_management ("flood")
Runtime error  Traceback (most recent call last):   File "<string>",
line 1, in <module>   File "c:\program files\arcgis\desktop10.1\arcpy
\arcpy\management.py", line 12937, in GetCount       raise e
ExecuteError: ERROR 000732: Input Rows: Dataset flood does not exist
or is not supported
>>> |
```

The layer flood does not exist in the current map document—it should have been floodzone. Press the UP ARROW key to access the previous line of code, as shown in the figure.

```
Python                                                              □ ×
>>> count = arcpy.GetCount_management ("flood")
Runtime error  Traceback (most recent call last):   File "<string>",
line 1, in <module>   File "c:\program files\arcgis\desktop10.1\arcpy
\arcpy\management.py", line 12937, in GetCount       raise e
ExecuteError: ERROR 000732: Input Rows: Dataset flood does not exist
or is not supported
>>> count = arcpy.GetCount_management ("flood")
```

Now the change can be made with little effort and the code can be run.

```
Python                                                            □ ×
>>> count = arcpy.GetCount_management ("flood")
Runtime error  Traceback (most recent call last):   File "<string>",
line 1, in <module>  File "c:\program files\arcgis\desktop10.1\arcpy
\arcpy\management.py", line 12937, in GetCount       raise e
ExecuteError: ERROR 000732: Input Rows: Dataset flood does not exist
or is not supported
>>> count = arcpy.GetCount_management ("floodzone")
>>>
```

The UP ARROW and DOWN ARROW keys limit the amount of retyping needed when you want to access previous lines of code.

The need for typing can also be minimized by using code Auto-completion prompts. For example, when you type `arcpy.Get`, you get a drop-down list of all the ArcPy functions that start with `Get`. You can select from the list by double-clicking an item or by scrolling through it, using the UP ARROW and DOWN ARROW keys. Once an item is selected, pressing the TAB key automatically completes the name of the function, as shown in the figure.

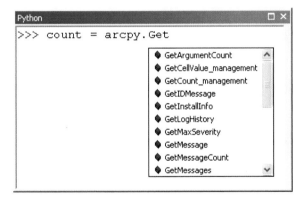

The Python window prompts also include aids for setting parameter values. For example, when you start entering tool parameters and an input feature class is expected, you are prompted by a list of available layers, like the example in the figure.

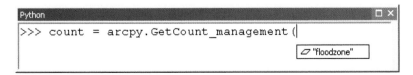

Prompts in the Python window also extend to variables defined in earlier lines of code. For example, when you start typing `print c`, you are prompted by a list of elements, starting with the letter `c`, that could logically be used. In this example, it includes the variable `count`, which was assigned a value in the previous line.

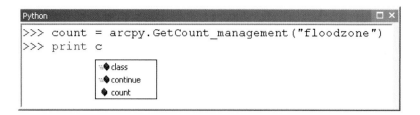

Making effective use of these prompts limits the amount of typing necessary, which serves to reduce typos. The prompts also remind you of the proper syntax, so they can help you learn how to write Python code.

## 3.5  Exploring Python window options

Several other options are available by right-clicking in the Python window.

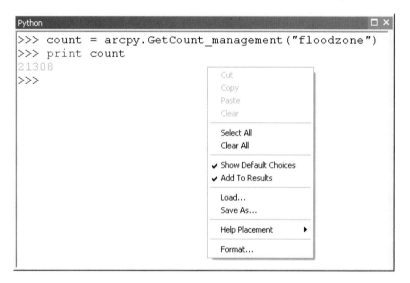

**>>> TIP**

Another way to save on typing is to drag tools and data directly into the Python window. With the Python window open, you can drag tools inside from ArcToolbox or data elements from the ArcMap Table Of Contents window or the Catalog window. Similar to the use of Autocompletion prompts, this not only saves time on typing, but it also reduces typos.

Among these options:

- *Cut, Copy, Paste,* and *Clear* provide basic editing of lines of code. Use your pointer to select part of the code to use these options.

- *Select All* allows you to select and then copy all lines of code in the Python window. This can be useful if you want to copy the code into another file, such as a script.

- *Clear All* allows you to remove all lines of code and start again with an empty Python window.

- *Show Default Choices* can be turned on and off. When selected, code autocompletion displays the available options for completing a line of code. Typically, code autocompletion is a good choice to have.

- *Add To Results* can be turned on and off. When this option is selected, any tools that are run in the Python window are added to the Results window. Whether you benefit from this is a matter of preference.

- *Load* can be used to load existing code from a script into the Python window.

- *Save As* allows you to save the code in the Python window to a text file (.txt) or Python file (.py).

- *Help Placement* provides options for where to place the Help section within the Python window relative to the Python prompt.

- *Format* provides access to formatting options, including font choices and color.

Of these options, Load and Save As are some of the most useful and are discussed in more detail in the following sections.

## 3.6 Saving your work

The Python window is great for running relatively short sections of Python code. You can experiment with Python statements and get immediate results. Single lines of code can also be easily run a second time. The Python window, however, is not intended as an environment to create longer and more complex sections of Python code. For this, you need Python scripts.

To benefit from the code you have already written in the Python window, you can save the contents for further use by right-clicking in the Python window and clicking Save As. The contents of the Python window can be saved as a text file (.txt) or a Python file (.py). The two formats are as follows:

1. *Text file:* This format saves everything that is visible in the Python window, including prompts and messages. It is like doing a Select All, and then copying and pasting the text into a text editor.

2. *Python file:* This format saves only Python code, which does not include prompts and messages.

Consider the example of Python code, shown in the figure.

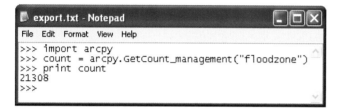

Saving the contents of the Python window to a text file results in saving all the text and symbols, as shown in Notepad in the figure.

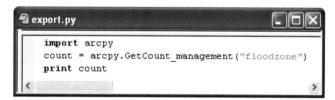

Saving the contents of the Python window to a Python file results in saving only the code, without the prompts and messages, as shown in PythonWin in the figure:

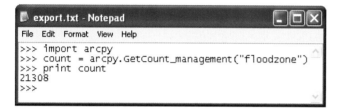

When you have a number of lines of Python code that work well and you want to save those lines for use in a script, you can simply copy and paste selected lines from the Python window into a Python editor. For longer sections of code, however, the Save As option gives you an option to save the code and remove all the noncode elements, such as prompts and messages.

# 3.7 Loading code into the Python window

The Python window is an interactive interpreter, and in general, you can use it for the quick execution of a few short lines of code. However, you can also load code that has already been written into the Python window. You can copy code from a Python editor and paste it into the Python window. You can also use the Load option to import the complete contents of a script.

Consider the script that was previously exported (export.py). Starting off with an empty Python window, right-click within the Python window and click Load. Then select a Python script file, such as the export.py script.

```
Python                                                    □ ×
>>> import arcpy
... count = arcpy.GetCount_management("floodzone")
... print count
...
```

When the existing script is loaded, notice that all the lines of code other than the first one are preceded by a secondary prompt. The secondary prompts indicate that the lines of code from the script have not been run—that is, they are not run one by one as they are loaded. This allows you to preview all the lines of code and make any necessary changes prior to running the code.

Normally, you would not want to run very lengthy scripts from within the Python window, but loading blocks of existing code saves time compared with typing or copying and pasting individual lines of code.

Being able to save your work to files, to load existing code from files, and to copy and paste code from existing scripts makes it easy to go back and forth between the Python window and a Python editor. Ultimately, you will typically want to save your final code as a script file for future use.

## Points to remember

- The Python window is a quick and convenient way to run geoprocessing tools while taking advantage of other Python modules and libraries. It also provides a great way for beginners to learn Python.

- You can save code from the Python window to a script file and you can load code from existing scripts into the Python window.

# Chapter 4
## Learning Python language fundamentals

## 4.1 Introduction

This chapter covers the fundamentals of the Python language, starting with the use of numbers and strings, variables, statements and expressions, functions, and modules. This makes up the basic syntax for writing code in Python. The second part of the chapter deals with controlling workflow, including the use of conditional statements, branching, and looping. These statements are an important element of geoprocessing scripts in ArcGIS and are fundamental to batch operations. The third part of the chapter is dedicated to best practices in writing scripts, including naming conventions and commenting on contents. PythonWin will be used as the editor to illustrate example scripts.

## 4.2 Locating Python documentation and resources

Before getting into Python syntax, it is useful to review where to look for Python documentation. If you installed Python as part of a typical ArcGIS for Desktop installation, you can find the Python manuals by clicking the Start button on the taskbar, and then, on the Start menu, clicking All Programs > ArcGIS > Python 2.7 > Python Manuals. The Python manuals include the complete documentation for the version of Python that is installed on your computer.

The same Python documentation can also be found online at http://docs.python.org. You will find multiple versions of the document for the various Python versions, including 2.6, 2.7, 3.1, 3.2, etc. You can also download the complete documentation in PDF, HTML, and text formats.

The Python documentation is quite extensive. The PDF version of the Python Library Reference, for example, is over 1,000 pages. This can be rather intimidating if you are just getting started in Python. The documentation, however, is not meant to be read cover to cover. Instead, the documentation is typically used to selectively look up the syntax for specific tasks. This chapter introduces you to the fundamentals of Python, which will give you the basic terminology needed to use the documentation more effectively.

The Python website at http://www.python.org also contains many additional resources to learn Python, including a "Beginner's Guide to Python" at http://wiki.python.org/moin/BeginnersGuide and a set of introductory tutorials at http://wiki.python.org/moin/BeginnersGuide/NonProgrammers. You will quickly realize that there is a wealth of resources on Python, created by a large and active user community

## 4.3 Working with data types and structures

Python uses a number of different *data types*, including strings, numbers, lists, tuples, dictionaries, and more. The data type of an object determines what type of values it can have and what operations can be performed on the object. *String* values consist of one or more characters, which can include letters, numbers, or other types of characters. There are two numeric data types: *integers* (whole numbers) and *floats*, or floating-point numbers (fractional numbers). Lists, tuples, and dictionaries are more complex data structures that consist of a collection of data elements.

In addition to data types, Python uses different *data structures*. A data structure is a collection of data elements that are structured in some way—for example, elements that are numbered in some way. The most basic data structure in Python is the *sequence*, in which each element is assigned a number, or index. Strings, lists, and tuples are examples of sequences. Because sequences share the same inherent data structure, there are certain things you can do with any type of sequence, and you will see examples of working with sequences later in this chapter.

Strings, numbers, and tuples are *immutable*, which means you can't modify them but only replace them with new values. Lists and dictionaries are *mutable*, which means the data elements can be modified. This chapter describes numbers, strings, and lists. Tuples and dictionaries are covered in chapter 6.

## 4.4 Working with numbers

Numbers can be integers or floats. Integers are whole numbers—that is, numbers that have no fractional part, meaning no decimals. For example, 1 and -34. Floats, or floating-point numbers, are numbers that have a decimal point. For example, 1.0 and -34.8307. Although both integers and floats are numeric data types, they act differently depending on Python syntax, so the distinction is important.

Consider a simple example:

```
>>> 3 * 8
```

The result is the value of 24. Simple enough. Another basic example:

```
>>> 16 / 4
```

The result is the value of 4. How about the next calculation:

```
>>> 17 % 4
```

The % operator here stands for *modulus,* or the remainder after the division. So, 17 % 4 evaluates the remainder of 17/4, which results in 1.

Next, take a look at a minor peculiarity in Python:

```
>>> 7 / 3
```

The code results in 2? What is going on here? When Python performs integer division, the result is always an integer. Any fractional part is ignored. If you want floating-point division, also referred to as *true division*, at least one of the inputs must be a floating-point number:

```
>>> 7 / 3.0
```

The code results in 2.33333333333.

The basic mathematical operators for integers and floats are summarized in table 4.1.

*Note: Starting in Python 3.0, all division will be true division so this peculiarity will no longer exist.*

**Table 4.1  Mathematical operators using integers and floating-point numbers**

| Operator | Description | Integer | | Floating-point | |
|---|---|---|---|---|---|
| | | Example | Result | Example | Result |
| * | Multiplication | 9 * 2 | 18 | 9 * 2.0 | 18.0 |
| / | Division | 9 / 2 | 4 | 9 / 2.0 | 4.5 |
| % | Modulus | 9 % 2 | 1 | 9 % 2.0 | 1.0 |
| + | Addition | 9 + 2 | 11 | 9 + 2.0 | 11.0 |
| - | Subtraction | 9 - 2 | 7 | 9 – 2.0 | 7.0 |

# 4.5 **Working with variables and naming**

Python scripts use *variables* to store information. A variable is basically a name that represents or refers to a value. For example, you may want to use a variable x to represent the number 17. Here is how you would do it in Python:

```
>>> x = 17
```

This is called an *assignment* statement. You assign the value of 17 to variable x. Once you assign a value to a variable, you can use the variable in an expression. For example:

```
>>> x = 17
>>> x * 2
34
```

Note that you need to assign a value to a variable before you can use it. So the line of code x * 2 requires that it be preceded by the line of code x = 17, which assigns the value of 17 to x.

A brief note about other programming languages is in order here. If you are familiar with languages like VBA or C++, you will have learned to first *declare* a variable and define its type (string, numeric, and others) before you can actually assign it a value. Python does not require you to declare it and you can directly assign it a value. If you have not used other programming languages, the use of variables in Python is fairly intuitive. It is also very efficient and results in fewer lines of code.

So how does Python know what type your variable is? It is implicit in how you assign values to it. For instance, x = 17 means that x is an integer, whereas x = 17.629 means that x is a float, and x = "GIS" means that x is a string. This is known as *dynamic* assignment. You can change the type of a variable by assigning it a new value.

Here are some basic rules for naming variables:

- Variable names can consist of letters, digits, and underscores (_).

- Variable names cannot begin with a digit. So var1 is a valid variable name, but 1var is not.

- Python keywords cannot be used as variable names. This includes statements such as print and import. Later in this chapter, you will see more keywords, as well as a way to view all the keywords in Python.

> **>>> TIP**
>
> You should use a single space around operators — for example, x = 17 is recommended, rather than x=17.

In addition to the rules for variables that are technically correct, here are some guidelines for creating good variable names:

- *Use descriptive names.* Thinking ahead about what you call a variable will help you write code that is easy to follow. It can also help you to remember what the variable means. For example, a variable called count is more meaningful than one simply called c.

- *Follow conventions.* Most programming languages have certain traditions, and Python has an official *Style Guide for Python Code*. The suggested style for variable names is to use short, all-lowercase names with words separated by underscores (_) as necessary for readability. Avoid using an underscore as the first letter, since these have special meaning in Python. The complete style guide can be found at http://www.python.org/dev/peps/pep-0008/.

- *Keep it short.* Although long variable names are technically OK, why call your variable something like number_of_cells_in_a_ raster_dataset? Lengthy names increase the chances of typos and make your code hard to read.

Naming Python scripts follows these same basic rules and guidelines.

### >>> TIP

Multiple variables can be assigned on the same line, which keeps your scripts compact. For example:

```
>>> x, y, z = 1, 2, 3
```

is the same as:

```
>>> x = 1
>>> y = 2
>>> z = 3
```

## 4.6 Writing statements and expressions

Once you have variables to work with, you can start writing Python *expressions* and *statements*.

An *expression* is a value—for example, 2 * 17 is an expression, representing the value of 34. Simple expressions are built from literal values (such as 17) by using operators and functions. More complicated expressions can be created by combining simpler expressions. Expressions can also contain variables.

A *statement*, on the other hand, is an instruction that tells the computer to *do* something. This can consist of changing the value of variables through assignment, printing values to the screen, importing modules, and many other operations.

The difference between expressions and statements is a bit subtle but important. Consider the following example:

```
>>> 2 * 17
34
```

Here, `2 * 17` is an expression. It has a value (34), which is automatically printed to the screen of the interactive Python interpreter. Now consider the following:

```
>>> x = 2 * 17
```

Here, `x = 2 * 17` is a statement. The variable `x` is assigned a value by the statement, but the statement itself is not a value. This is a defining property of statements: *Statements do something.* A statement itself is not a value, and hence, in the preceding example, the interactive Python interpreter does not print a value—you would need to use the `print` statement for that:

```
>>> x = 2 * 17
>>> print x
34
```

Statements that assign values to variables are among the most important statements in any programming language. At first glance, assignment statements may appear to serve only as temporary containers for values, but the real power lies in the fact that you do not need to know what values they hold to be able to manipulate them. So you can write Python scripts that manipulate variables without knowing the values they may eventually hold.

## 4.7  Using strings

A second important data type is strings. A set of characters surrounded by quotation marks is called a *string literal*. You already saw an example of it when you wrote the code `print "Hello World"`. You can create a string variable by assigning a string literal to it.

You can use single quotation marks (' ') or double quotation marks (" ")—in Python—they serve the same purpose. Quotation marks are like bookends, telling the computer where the string begins and where it ends. Having two ways to indicate bookends provides flexibility. For example, if you use a pair of double quotation marks to surround your text, you can use as many single quotation marks inside the string as you want, and vice versa. For example:

```
>>> print "I said: 'Let's go!'"
```

Using single quotation marks around this string would be confusing, and it would also generate a syntax error.

Strings are used frequently in geoprocessing scripts, in particular when defining geoprocessing tool parameters. For example, paths or partial paths are used to define the inputs and outputs of a tool and these paths are stored as strings. Therefore, string operators become very important in working with datasets in different workspaces.

Consider a few things you can do with strings. For example, you can concatenate strings by simply adding them up:

```
>>> x = "G"
>>> y = "I"
>>> z = "S"
>>> print x + y + z
GIS
```

When combining strings to form a new string, you may want to add spacing in between by using double quotation marks around a space (" ")—for example:

```
>>> x = "Geographic"
>>> y = "Information"
>>> z = "Systems"
>>> print x + " " + y + " " + z
Geographic Information Systems
```

Strings can contain numeric characters. However, when trying to combine strings and numbers, numbers need to first be converted to a string. Consider the following example:

```
>>> temp = 32
>>> print "The temperature is " + temp + " degrees"
```

The result is an error message because you cannot add a string and a number together. You can use the notation `str` to convert the number to a string, however. The correct code is as follows:

```
>>> temp = 100
>>> print "The temperature is " + str(temp) + " degrees"
The temperature is 100 degrees
```

In this example, `str` is an example of a *function*, which is covered later in this chapter. Converting the value of the variable from one type to another is known as *casting*. The preceding example code is a simple example of string formatting using the + operator to concatenate strings. Alternative ways to format strings are covered later in this chapter.

# 4.8 **Using lists**

A third important data type is *lists*. A Python list of items is surrounded by square brackets ( [ ] ), and the items are separated by commas ( , ). The items themselves can consist of numbers, strings, or several other data types.

Lists are used frequently in geoprocessing scripts. For example, you may want to create a list of all the feature classes in a workspace and perform an operation on all the feature classes in the list.

You can create a list by typing all the items. The following example creates a list of numbers:

```
>>> mylist = [1, 2, 4, 8, 16, 32, 64]
```

Notice that the items are separated by a comma ( , ), followed by a space. The space is not required but makes the list much easier to read.

List items are not limited to numbers, but can also consist of strings:

```
>>> mywords = ["jpg", "bmp", "tif", "img"]
```

You can print the contents in the list using a `print` statement as follows:

```
>>> print mywords
["jpg", "bmp", "tif", "img"]
```

Notice that the items retain their original order when printed. A list is an ordered set of items. Later in this chapter, you will see how to manipulate lists.

# 4.9 **Working with Python objects**

Now that you have seen a number of data types in Python (for example, numbers, strings, and lists), it is important to revisit the concept of Python being an object-oriented programming language and of everything in Python being an object. Each object has (1) a value; (2) an identity, which is a unique identifier; and (3) a type, which is the kind of values an object can hold.

Consider the following example:

```
>>> name = "Paul"
```

This statement creates an object (name), and this object has a value (Paul):

```
>>> name
'Paul'
```

The object also has a unique identifier, which varies depending on the specific computer being used:

```
>>> id(name)
593835200
```

The unique identifier is provided by the computer to keep track of the object (and its value and properties), but it is not important to know the actual number being used.

Finally, the object has a type:

```
>>> type(name)
<type 'str'>
```

A second important concept is that variables in Python are dynamic. Consider the following example:

```
>>> var1 = 100
>>> type(var1)
<type 'int'>
>>> var2 = 2.0
>>> type(var2)
<type 'float'>
```

The object type of the variables is determined by the nature of the value assigned to it. This is somewhat unique in Python, because in many other programming languages, variables first need to be declared (and given a type).

Object type conversion can be accomplished using casting. Consider the following example:

```
>>> var = 100
>>> newvar = str(var)
>>> type(newvar)
<type 'str'>
```

In the preceding example, the second line of code casts the variable var as a string—the value is still 100, but now it is a string, not an integer. Trying to cast a string such as "Paul" as a number, however, is not logical and therefore would result in an error. Casting an integer as a float, or vice versa, on the other hand, is very common.

# 4.10 **Using functions**

Python expressions and statements use variables and functions. Variables are discussed earlier in the chapter. A function is like a little program that is used to carry out a certain action. There is a set of core functions in Python, referred to as the built-in functions, which you can use directly in any statement. An example of the power function `pow` follows:

```
>>> pow(2,3)
8
```

The `pow` function returns 2 to the power of 3, or the value of 8. Using a function is referred to as *calling* the function. When you call a function, you supply it with *parameters* (in this case, 2 and 3), and it returns a value. And because it returns a value, a function call is a type of expression.

To see which built-in functions are available, you can consult the Python manuals, but you can also access them directly within Python itself using the `dir(__builtins__)` statement:

```
>>> print dir(__builtins__)
```

*Note: There are two underscores (__) preceding and following "builtins."*

This code prints a list of several dozen built-in functions—too many to review here in detail, but several of them are used in upcoming examples. Once you identify a function you may want to use, you need to review the description of the function and the syntax. To see the details of a particular function, use the `__doc__` statement. For example:

```
>>> print pow.__doc__
```

*Note: There are two underscores (__) preceding and following "doc."*

The print statement is not required in the interactive Python interpreter, but it provides better formatting of the output. Note the description of the pow function:

```
pow(x, y[, z]) -> number. With two arguments, equivalent to x**y.
With three arguments, equivalent to (x**y) % z, but may be more efficient
(e.g. for longs).
```

Notice that the `pow` function has three parameters, also referred to as *arguments*, which are separated by commas (,). Two of these arguments are required (x and y) and the third is optional, which is indicated by the use of square brackets ([ ]). Also notice that the description refers to (**), which is the basic operator in Python for exponentiation.

Several other common built-in functions are listed in table 4.2. This list is not exhaustive, but it contains some of the more widely used functions.

**Table 4.2  Common built-in Python numeric functions**

| Function | Description | Example | Returns |
|----------|-------------|---------|---------|
| abs(x) | Returns the absolute value of a number | abs(-8) | 8 |
| float(x) | Converts a string or a number to a float | float("8.0") | 8.0 |
| int(x) | Converts a string or a number to an integer | int("8") | 8 |
| pow(x,y[,z]) | Returns x to the power of y | pow(4,3) | 64 |
| round(x[,n]) | Rounds off the given float to n digits after the decimal | round(2.36) | 2 |
| str(x) | Returns a string representation of a given object | str(10) | "10" |

Python is not limited to built-in functions, and additional functions can be accessed using modules, as discussed later in this chapter. You can also create your own functions, which is covered in chapter 12.

# 4.11  Using methods

Methods are similar to functions. A method is a function that is closely coupled to an object—for example, a number, a string, or a list. A method is called as follows:

```
<object>.<method>(<arguments>)
```

This looks just like calling a function, but now the object is put before the method with a dot (.) separating them. Here is a simple example:

```
>>> topic = "Geographic Information Systems"
>>> topic.count("i")
2
```

In this example, the variable *topic* is assigned a string. Because the variable is defined as a string, Python automatically makes all methods associated with a string object available. In the second line of code, the `count` method is called. The argument, in this case, is a substring (the letter *i*), and the result is a count of the number of times the substring occurs in the string object. The `count` method is case sensitive, as is most of Python syntax.

Methods are not limited to strings—many of Python's built-in data types have associated methods, and they are widely used when working with objects in Python scripts.

# 4.12 **Working with strings**

Strings are a useful built-in data type and are frequently used in Python scripts. Many of the geoprocessing variables are string types—for example, the names of map documents in a workspace, the names of feature classes in a geodatabase, and the names of fields in a table. In many cases, these strings are somewhat complex. For example, the path of a feature class might look like this: C:\EsriPress\Python\Data\Exercise03\zipcodes.shp. It is therefore useful to review a number of additional string methods.

The `lower` method returns a lowercase version of the string value:

```
>>> mytext = "GIS is cool"
>>> print mytext.lower()
gis is cool
```

The `upper` method returns an uppercase version of the string value:

```
>>> mytext = "GIS is cool"
>>> print mytext.upper()
GIS IS COOL
```

The `title` method returns a title-cased version of the string value—that is, each word starts with an uppercase letter:

```
>>> mytext = "GIS is cool"
>>> print mytext.title()
Gis Is Cool
```

Strings (and other Python sequences) have an index positioning system, which uses a range of values enclosed in square brackets ([ ]). Each character in a string is assigned an index value, starting with the index number of zero (0). Spaces are counted like any other characters. Consider the following string:

```
>>> mystring = "Geographic Information Systems"
```

The code to obtain the first character would look like this:

```
>>> mystring[0]
'G'
```

This approach can be used to obtain any character:

```
>>> mystring[23]
'S'
```

You can use negative index numbers to start counting from the end. The last item in the list is index -1. This makes it possible to get the last item without knowing the exact count:

```
>>> mystring[-1]
's'
```

Strings can be sliced into smaller strings. *Slicing* uses two indices separated by a colon (:). The first index is the number of the first character you want to include. The second index is the number of the first character you do not want to include. For example, the following code creates a new list containing the elements starting with index number 11, up to but not including the element with index number 22:

```
>>> mystring = "Geographic Information Systems"
>>> mystring[11:22]
'Information'
```

Leaving out one of the indices means you are not putting a limit on the range. For example, the following code creates a new string consisting of the characters starting with index number 11, up to and including the highest index number:

```
>>> mystring = "Geographic Information Systems"
>>> mystring[11:]
'Information Systems'
```

The find method can be used to identify a substring and returns the leftmost index when the string is found. A value of -1 is returned when the string is not found.

```
>>> mystring = "Geographic Information Systems"
>>> mystring.find("Info")
11
```

The Python find method is case sensitive:

```
>>> mystring = "Geographic Information Systems"
>>> mystring.find("info")
-1
```

The returned value of -1 means that the substring was not found.

The in operator is similar to the find method but returns a Boolean value:

```
>>> mystring = "Geographic Information Systems"
>>> "Info" in mystring
True
```

The join method is used to join elements of a list:

```
>>> list_gis = ["Geographic", "Information", "Systems"]
>>> string_gis = " "
>>> string_gis.join(list_gis)
'Geographic Information Systems'
```

Here, the elements in the list are joined into a single string. The string object (string_gis) has a method called join. The value of the string object, in this case, is a space (" ") and the argument of the join method is the list of elements to be joined into a single string.

The opposite of the join method is the split method. The argument of the split method is the separator to be used to split the input string into elements—also a space (" ").

```
>>> pythonstring = "Geoprocessing using Python scripts"
>>> pythonlist = pythonstring.split(" ")
>>> pythonlist
['Geoprocessing', 'using', 'Python', 'scripts']
```

Another commonly used method to manipulate path and file names is the strip method. The generic strip method allows you to remove any combination of characters in any order from the ends of an existing string. For example:

```
>>> mytext = "Commenting scripts is good"
>>> mytext.strip("Cdo")
'mmenting scripts is g'
```

Notice that the strip method is not very specific: It removes any characters from the start or the end of the string, irrespective of the order in which the characters are listed in the argument, and irrespective of whether all the characters are included on either end of the string.

The lstrip and rstrip methods provide a bit more control by limiting stripping to either the left side or the right side of the string, respectively. For example:

```
>>> mytext = "Commenting scripts is good"
>>> mytext.rstrip("Cdo")
'Commenting scripts is g'
```

Notice that, in this case, the leading "Co" is not removed because the `rstrip` method removes only trailing characters.

Calling one of the strip methods without any arguments removes spaces, also referred to as *whitespace* in coding. This is useful for cleaning up strings that may have been formatted in another program.

The `replace` method is much more specific and replaces all occurrences of a specific substring by another substring. This works just like a find-and-replace operation in a text editor. For example:

```
>>> mygis = "Geographic Information Systems"
>>> mygis.replace("Systems", "Science")
'Geographic Information Science'
```

*Note: Be very careful using stripping methods, particularly the strip method itself, because the order in which characters are entered as arguments does not matter. It is easy to remove more characters than you intended.*

The `replace` method could also be used to remove the file extensions from file names by replacing a particular substring with an empty string (""). This approach is more specific than the `strip` method, because it removes a specific substring wherever it occurs, and then replaces it. For example:

```
>>> myfile = "streams.shp"
>>> myfile.replace(".shp", "")
'streams'
```

The `format` method is commonly used for string formatting. Its most basic usage is to insert a value into a string using a single placeholder. Consider the following example:

```
>>> temp = 100
>>> print "The temperature is {0} degrees".format(temp)
The temperature is 100 degrees
```

In this example, {0} is a *replacement field*, which is replaced by the argument of the `format` method—in this case, `temp`. This approach to string formatting is not limited to single replacement fields:

```
>>> username = "Paul"
>>> password = "surf$&9*"
>>> print "{0}'s password is {1}".format(username, password)
Paul's password is surf$&9*
```

*Note: String formatting can also be accomplished using the % operator, and you will likely encounter this in existing scripts. The format method, however, is the recommended approach to string formatting.*

## >>> TIP

In Python, you may encounter Unicode strings, which look like regular strings preceded by the letter *u*. For example: `u'roads.shp'`. In general, you can think of strings as "plain text." Text is stored as particular characters, and different languages use different sets of characters. This can cause problems across different computer platforms, and characters may not display correctly. You may have encountered this when trying to read web pages or e-mails in different languages. To overcome these problems, the Unicode system is designed to represent every character from every language. In general, Unicode strings work just like regular strings but are more robust when working with different international sets of characters.

## 4.13 Working with lists

Lists are a versatile Python type and can be manipulated in many different ways. In the previous section, you saw how items in a list were joined to form a single string and how a single string was split into items in a list. In this section, you will see a few more ways to manipulate lists.

Consider the following list:

```
>>> cities = ["Austin", "Baltimore", "Cleveland", "Denver", "Eugene"]
```

The number of items in a list can be determined using Python's built-in `len` function, as follows:

```
>>> print len(cities)
5
```

Lists can be sorted using the `sort` method. The default sorting is alphanumerical, but it can be reversed by using the `reverse` argument of the `sort` method, as follows:

```
>>> cities.sort(reverse = True)
>>> print cities
['Eugene', 'Denver', 'Cleveland', 'Baltimore', 'Austin']
>>> cities.sort
>>> print cities
['Austin', 'Baltimore', 'Cleveland', 'Denver', 'Eugene']
```

Just like strings, Python lists are indexed, starting with the index number zero (0). These index numbers can be used to obtain specific items in the list or to slice the list into smaller lists. The code to obtain the second item from the preceding list would look as follows:

```
>>> cities[1]
'Baltimore'
```

You can use negative index numbers to start counting from the end. The last item in the list is index -1. This makes it possible to get the last item without knowing the exact count:

```
>>> cities[-1]
'Eugene'
```

The following code can be used to obtain the second to the last item:

```
>>> cities[-2]
'Denver'
```

Lists can be sliced into smaller lists. Slicing uses two indices separated by a colon (:). The first index value is the number of the first element you want to include. The second index is the number of the first element you do not want to include. For example, the following code creates a new list containing the elements starting with index number 2, up to but not including the element with index number 4:

```
>>> cities[2:4]
['Cleveland', 'Denver']
```

*Note: Using a single index number, for example, Cities[1], returns the value of a single item—in this case, a string. Slicing returns a new list, even if it contains only a single item.*

Leaving out one of the indices means you are not putting a limit on the range. For example, the following code creates a new list consisting of the items starting with index number 2, up to and including the highest index number:

```
>>> cities[2:]
['Cleveland', 'Denver', 'Eugene']
```

The following code can be used to obtain the items up to but not including the index number:

```
>>> cities[:2]
['Austin', 'Baltimore']
```

Another important list operation determines membership using the `in` operator. It checks whether something is true and returns a value of `True` or `False`. Consider the following example:

```
>>>cities = ["Austin", "Baltimore", "Cleveland", "Denver", "Eugene"]
>>> "Baltimore" in cities
True
>>> "Seattle" in cities
False
```

Elements can be deleted using the `del` statement. For example, the following code deletes a specific element from a list based on the index number:

```
>>> cities = ["Austin", "Baltimore", "Cleveland", "Denver", "Eugene"]
>>> del cities[2]
>>> cities
['Austin', 'Baltimore', 'Denver', 'Eugene']
```

In addition to carrying out operations on lists, you can manipulate lists using methods. List methods include `append`, `count`, `extend`, `index`, `insert`, `pop`, `remove`, `reverse`, and `sort`. The last two list methods have already been mentioned in this section. A brief discussion of some of the other list methods follows.

The `append` method can be used to append an element to the end of the list:

```
>>> cities = ["Austin", "Baltimore"]
>>> cities.append("Cleveland")
>>> cities
['Austin', 'Baltimore', 'Cleveland']
```

The `count` method determines the number of times an element occurs in a list:

```
>>> yesno = ["True", "True", "False", "True", "False"]
>>> yesno.count("True")
3
```

The `extend` method allows you to append several values at once:

```
>>> list1 = [1, 2, 3, 4]
>>> list2 = [11, 12, 13, 14]
>>> list1.extend(list2)
>>> list1
[1, 2, 3, 4, 11, 12, 13, 14]
```

The `index` method is used to find the index of the first occurrence of a value:

```
>>> mylist = ["The", "quick", "fox", "jumps", "over", "the", "lazy", ➜
➜ "dog"]
>>> mylist.index("the")
5
```

**Note:** *This book uses an arrow symbol* ➜ *to indicate long lines of code that appear all on one line in Python.*

The `insert` method makes it possible to insert an element into a list at a particular location:

```
>>> cities = ["Austin", "Cleveland", "Denver", "Eugene"]
>>> cities.insert(1, "Baltimore")
>>> cities
['Austin', 'Baltimore', 'Cleveland', 'Denver', 'Eugene']
```

The `pop` method removes an element from a list at a particular location and returns the value of this element:

```
>>> cities = ["Austin", "Baltimore", "Cleveland", "Denver", "Eugene"]
>>> cities.pop(3)
'Denver'
>>> cities
['Austin', 'Baltimore', 'Cleveland', 'Eugene']
```

The `remove` method is used to remove the first occurrence of a value:

```
>>> numbers = [1, 0, 1, 0, 1, 0, 1, 0, 1, 0]
>>> numbers.remove(0)
>>> numbers
[0, 1, 0, 1, 0, 1, 0, 1, 0]
```

There is no need to memorize all these methods. Code autocompletion prompts will help you by listing all the applicable methods. For example, consider the following code example in the Python window. After you have created the list called cities, the interactive Python interpreter recognizes it as a list when you start typing the next line of code. So after typing `cities` followed by a dot (.), you are prompted by a drop-down list of list methods, like the example in the figure.

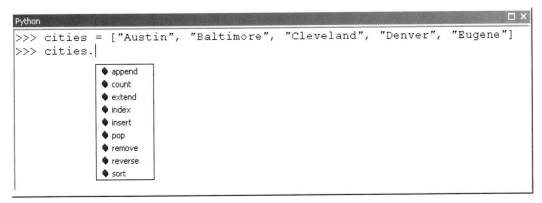

You will see more examples of working with lists in chapter 6.

# 4.14 **Working with paths**

When files are organized in a computer, a folder structure is used to facilitate retrieving files using a path. A path consists of a list of folder names separated by a backslash (\), optionally followed by a file name. An example of a path for a workspace is C:\EsriPress\Python\Data\Exercise04. An example of a path for a shapefile is C:\EsriPress\Python\Data\Exercise02\rivers.shp.

Notice that the backslash (\) is commonly used when writing paths. Python, however, treats a backslash as an escape character. For example, \n represents a line feed, and \t represents a tab. This means that in Python, you should avoid using backslashes in paths. There are three correct ways to specify a path in Python:

1.  Use a forward slash (/)—for example, `"C:/EsriPress/Python/Data"`.

2.  Use two backslashes (\\)—for example, `"C:\\EsriPress\\Python\\Data"`.

3.  Use a string literal by placing the letter *r* before a string—for example, `r"C:\EsriPress\Python\Data"`. The letter *r* stands for "raw string," which means that a backslash will not be read as an escape character.

The style you use for paths is a matter of preference. It is recommended to adopt a single style and stay with it. This book uses forward slashes, such as `"C:/EsriPress/Python/Data"`, but it is good to be aware of the other types of notation.

In Python, paths are stored as strings. For example, the following code assigns the path for a shapefile to a string variable:

```
>>> inputfc = "C:/EsriPress/Python/Data/Exercise02/rivers.shp"
```

Once a path is assigned to a variable, the path can be used to reference data on disk.

> **Note:** *When not referencing code, this book uses the regular backslash (\) for paths, such as C:\EsriPress\Python\Data\Exercise02\rivers.shp.*

# 4.15 **Working with modules**

There are many more functions available in Python than just the built-in functions. Using them requires the use of *modules*. Modules are like extensions that can be imported into Python to extend its capabilities. Typically, a module consists of a number of specialized functions. Modules are imported using a special statement called `import`. A commonly used module is the `math` module. Importing it into Python works like this:

```
>>> import math
```

Once you import a module, all functions in that module are available to use in Python. To call a function from the imported module, you must still refer to the module itself by writing `<module>.<function>`. For example, to use the `cosine` function in the `math` module, you would use the following code:

```
>>> math.cos(1)
0.54030230586813977
```

Note that the `math.cos` function assumes the input value is in radians. To get the description of a function, use the `__doc__` statement:

```
>>> print math.cos.__doc__
```

Note the result:

```
cos(x): Return the cosine of x (measured in radians)
```

To get the list of all the functions in the `math` module, use the `dir` statement:

```
>>> dir(math)
```

One of the reasons for using the `<module>.<function>` syntax to call a function is that functions from different modules can have the same name. If you are sure you won't import more than one function with the same name in the same script, you can shorten your code by using a variant of the `import` statement:

```
>>> from math import cos
>>> cos(1)
0.54030230586813977
```

Once you use the `from <module> import <function>` statement, you can use the function without its module prefix.

Another module is the `time` module. You can review the functions of the `time` module using the following code:

```
>>> import time
>>> print dir(time)
```

The code produces a list of all the functions in the `time` module. Try out some of the simpler ones. The `time.time` function determines the current time as the number of seconds since the "epoch," or reference, date. It is platform-dependent—for UNIX, it is 0 hours January 1 1970, and that is what Python uses by default.

```
>>> time.time()
1295104277.9679999
```

The result varies from platform to platform, but you can use the function to reliably time something. For example, if you wanted to determine how long it takes to carry out a certain procedure, you would record the time before and after the procedure to determine the amount of time.

The `localtime` function converts the time in seconds to the components that make up the current local time and date:

```
>>> time.localtime()
time.struct_time(tm_year = 2011, tm_mon = 1, tm_mday = 15, tm_hour = 8, →
→ tm_min = 40, tm_sec = 35, tm_wday = 5, tm_yday = 15, tm_isdst = 0)
```

The code returns a structure known as a `tuple` (rhymes with quintuple), which contains the following elements: year, month, day of the month, hour, minute, second, day of the week (Monday is 0), day of the year, and daylight saving time. Tuples are covered in more detail in later chapters.

The `asctime` function converts time to a string:

```
>>> time.asctime()
'Sat Jan 15 08:44:05 2011'
```

Python contains many keywords, which should never be used as variable names. To see a list of keywords, use the `keyword` module:

```
>>> import keyword
>>> print keyword.kwlist
```

Examine the list and notice that the `print` statement is included, and so is `import`. You will become familiar with many of these keywords—for example, the statements `if`, `elif`, and `else` are covered in the section on workflows, later in this chapter.

Many more modules are available in Python, and you will see examples of them throughout the upcoming chapters. It is also important to recognize that you can import modules from other programs. The most relevant example is the ArcPy module from ArcGIS—ArcPy is actually referred to as a site package because it contains multiple modules. When you write scripts to work with ArcGIS, the first thing you typically do is to import ArcPy from ArcGIS for access to all the ArcGIS tools. ArcPy uses the same `<module>.<function>` syntax you have already seen in this chapter. Chapter 5 covers this syntax in greater detail.

## 4.16 Controlling workflow using conditional statements

The code you have looked at so far has a simple, sequential flow. Each statement or expression is run once, in the order of occurrence. More complex applications require that you be able to selectively run certain portions of your code or repeat parts of it. *Branching* is one way to control the workflow in your script. It basically means making a decision to take one path or another. Branching typically uses the `if` structure and its variants. Under the `if` structure, scripts can branch into a section of code or skip it, based on the conditions that are set. Consider a simple example:

```
import random
x = random.randint(0,6)
print x
if x == 6:
    print "You win!"
```

The `random.randint(0,6)` expression creates a random integer between 1 and 6, as if throwing a single die, and in PythonWin, this value gets printed to the Interactive Window. If the value is 6, the string "You win!" is printed to the Interactive Window. What happens if the value is not equal to 6? Nothing—the next line of code is skipped.

All `if` structures have a condition that is either true or false. Python has its own built-in values to represent true and false. `True` (with a capital T) represents true and `False` (with a capital F) represents false. This may seem very intuitive, but in earlier versions of Python, it was common to use 1 for true and 0 for false.

Conditions are most often created by using *comparison operators*. The basic operators are listed in table 4.3. Notice that the symbol for "equal to" is a double equals sign (==), not a single equals sign (=), as you might expect. The use of the single equals sign is reserved for assigning a value to a variable. So x = 6 is an assignment, and x == 6 is a condition.

**Table 4.3  Comparison operators**

| Operator | Description | Example Conditions | Result |
|---|---|---|---|
| == | Equal to | 4 == 9 | False |
| != | Not equal to | 4 != 9 | True |
| > | Greater than | 4 > 9 | False |
| < | Less than | 4 < 9 | True |
| >= | Greater than or equal to | 3 >= 3 | True |
| <= | Less than or equal to | 3 <= 2 | False |

Also notice a few things about the syntax of the branching structure. First, the if statement is followed by a colon (:)—this is required and indicates the beginning of indented code on the next line. Second, the if statement can be used on its own—that is, without being followed by an else or endif statement, as is often required in other programming languages. Third, the line following the if statement is indented—this is a critical part of coding in Python.

By indenting a line, the code becomes a *block*. A block is one or more consecutive lines indented by the same amount. Indenting sets lines off visually—and also logically. Blocks are commonly used as part of a branching structure. Blocks form the statement, or group of statements, that is run if the condition is True. Indenting in Python is not optional. It is the only way to define a block.

To review, the basic structure of an if statement is as follows:

```
if x == 6:
    print "You win!"
```

In the first line of code, the keyword if is followed by a condition and then a colon (:). Then there is a block of code—that is, one or more lines indented in the same manner. If the condition evaluates to True, the statements that make up the block are run. If the condition evaluates to False, the lines of code are skipped and the program moves on to the next statement after the if structure. Notice that there is no endif statement as you might expect if you are familiar with other programming languages. So how does Python know when you have reached the end of the if structure? When you stop indenting your code—so accurate indenting is the key to making this structure work.

>>> **TIP**
You can create indentation using tabs or spaces. There is some debate in the Python community as to which one is better—and how many spaces to use if you use spaces—but this is largely a matter of personal preference. The key is consistency—that is, if you indent blocks using four spaces, always use four spaces. Mixing tabs and spaces may appear to be identical visually, but it will cause problems. Common styles are to use one tab, two spaces, or four spaces. The choice is yours—just be consistent.

There are a number of variations on the `if` structure using the `elif` and `else` statements. In the following example, the `elif` statement is executed only if the condition in the `if` statement is `False`. The `elif` statement can be repeated as many times as necessary, giving you the option to specify an action for every possible input. The `else` statement is executed only if all the previous conditions are `False` and can be used only once in a single `if` structure. The `else` statement (if used) comes after all the `elif` statements and does not include a condition—it is executed only when none of the previous conditions evaluated to `True` and it does not test any additional condition. For example:

```
import random
x = random.randint(0,6)
print x
if x == 6:
    print "You win!"
elif x == 5:
    print "Try again!"
else:
    print "You lose!"
```

The `if` structure and its variants are referred to as *branching* structures, because they allow your code to branch into various directions depending on a condition. Under the `if` structure, only certain parts of your code are executed and other parts are skipped.

## 4.17 Controlling workflow using loop structures

Another way to control your workflow is to use *loop* structures. This allows you to repeat a certain part of your code until a particular condition is reached or until all possible inputs are used. There are two basic forms of loop structures in Python: `while` loops and `for` loops.

Here is an example of a simple `while` loop:

```
i = 0
while i <= 10:
    print i
    i += 1
```

The counter variable `i` is set to a beginning value of zero (0). The `while` statement checks the value of the counter variable. If this condition evaluates to `True`, the block of code is run. In this block, the value of the counter variable is printed, and on the next line, the value is increased by 1. After one iteration of the loop, the value of the counter variable is therefore 1.

The block of code is repeated until the condition evaluates to `False`. The result of the code is a printed list of the numbers 0 through 10.

`While` statements require an exit condition. The variable used in the exit condition (i, in the case of the preceding code) is called a *sentry variable*. The sentry variable is compared to some other value or values. It is important to make sure that your exit condition is robust—that is, after a certain number of repetitions, the exit condition must be reached; otherwise, the loop would keep going. For example, what if your script read as follows:

```
i = 0
while i <= 10:
    print i
```

In this case, the sentry variable does not change in the `while` loop and the exit condition is never reached. This results in an infinite loop. You definitely want to avoid infinite loops in your scripts because it typically means that you have to crash the application to exit. So be sure to confirm that the exit condition is, in fact, reached at some point.

There are a number of options to exit an infinite loop, however. When using the Python prompt, you can press CTRL + C. When using the PythonWin editor, you can right-click the PythonWin icon in the notification area, at the far-right corner of the taskbar, and click "Break into running code."

> **Note:** *If breaking out of an infinite loop using the preceding suggestions does not work, you can terminate the application that is running Python. On the Windows platform, press CTRL + ALT + DELETE. This brings up the Windows Task Manager, which allows you to end a specific application—for example, PythonWin or ArcMap if you are running code from the Python window.*

On the other hand, you also want to make sure it is possible for the exit condition to evaluate to `True` at some point—otherwise, the block will never run. Take the following code, for example:

```
i = 12
while i <= 10:
    print i
    i +=1
```

In this script, the code block will never run, because the value assigned to the sentry variable (12) prevents the condition in the `while` loop from ever evaluating to `True`. This may appear like simple logic, but as your scripts become more complex, it becomes easier to overlook the simple things.

The `while` loop repeats part of your code based on a condition. The `for` loop, on the other hand, also repeats part of your code, but not based on a condition. Instead, the `for` loop is based on a sequence—an ordered list of things. A `for` loop repeats a block of code for each element of the sequence. When it reaches the end of the sequence, the loop ends.

Here is an example of a simple `for` loop, as follows:

```
mylist = ["A", "B", "C", "D"]
for letter in mylist:
    print letter
```

In this example, `letter` is the name of a variable, and for each iteration of the loop, this variable is assigned a different value.

The result is a print of every value in the sequence. So a `for` loop iterates over every value in the sequence until the sequence is empty. The preceding example code uses a *list*, which is one of Python's data types (or data structures) to store sequences. A list uses square brackets ([ ]) around a sequence of elements that are separated by commas (,). Other data types for sequences are tuples and dictionaries. These data types are covered in chapter 6.

# 4.18 Getting user input

Many Python scripts require inputs from outside the script itself. There are a number of ways to obtain these inputs. The first one is to use a system argument, `sys.argv`. The Python `sys` module provides access to some variables used or maintained by the interactive Python interpreter and to functions that interact with the interpreter. Arguments can be passed to a script from within the interactive interpreter. For example, the multiply.py script in the figure includes two arguments that need to be provided by the user. ➔

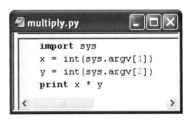

When the script is run, the arguments need to be specified on the Run Script dialog box. Arguments are separated by a space. Arguments typically consist of values or paths to files. ➔

Arguments are read as strings by default, so the code uses the `int` function to convert the values to strings. Once the script is run in PythonWin, the result is written to the Interactive Window.

System arguments passed to the script start at index number 1 because `sys.argv[0]` is the location of the script itself. The `sys.argv` method is widely used in Python scripts to receive information from the application that is calling the script. In later chapters, however, you will see a more robust alternative, which is commonly used when calling scripts from an ArcGIS for Desktop application.

The second way to obtain user input is to use the `input` function:

```
>>> x = input("")
```

Running this line of code in the interactive Python interpreter brings up a dialog box for you to enter a value in the text box, as shown in the figure.

When you enter a value and click OK, this input can now be used in the script.

*Note: The* input *function does not apply to working in the Python window in an ArcGIS for Desktop application, but only to PythonWin. Different Python editors vary in terms of how user input can be obtained using the* input *function.*

## 4.19 **Commenting scripts**

Well-developed scripts contain comments that help users understand them. Take a look at the AddressErrors.py script, as shown in the figure. The first section provides the name of the author, the version of the script, a brief description, and the license level required. In this case, the tool was authored by Esri staff. Comments are preceded by the number sign (#). When the script is run, any line that starts with the number sign is not executed.

```
AddressErrors.py

# Author: ESRI
# Date:    June 2010
#
# Purpose: This script checks street centreline data for errors in dual-range address attributes.
#          Errors reported are:
#
#              OVERLAP    - the address range overlaps the next segment
#              UNDERLAP   - the address range has a gap between the next segment
#              DIRECTION  - the segment range direction is opposite to the range origin
#              FROMTO     - the segment has a flipped from/to range
#              LEFTRIGHT  - the address ranges are on the wrong side
#              PARITY     - the address range disagrees with the assigned parity
#
#          Requires ArcGIS 10 - ArcInfo.
#
#
try:
    import arcpy
    import math
    import os
    import sys
    import traceback

    arcpy.env.overwriteOutput = True

    #Get the input feature class or layer
    inFeatures = arcpy.GetParameterAsText(0)
    inDesc = arcpy.Describe(inFeatures)
    if inDesc.dataType == "FeatureClass":
        inFeatures = arcpy.MakeFeatureLayer_management(inFeatures)
    searchRadius = str(inDesc.SpatialReference.XYTolerance * 10) + " " + \
                   str(inDesc.SpatialReference.LinearUnitName).replace('Foot_US','Feet')
    xyTol = inDesc.SpatialReference.XYTolerance
    inPath = os.path.dirname(inDesc.CatalogPath)
    sR = inDesc.spatialReference
    rangesAreText = False
```

Comments can also be placed on the same line after pieces of code. For example, later in the same `AddressErrors.py` script, there is a section that contains lines embedded with comments preceded by the number sign, as shown in the figure.

```
AddressErrors.py
        for f in inputFields:
            if len(f) > 0: #parity fields may be empty
                fMap = arcpy.FieldMap()
                fMap.addInputField(inFeatures,f)
                fieldMappings.addFieldMap(fMap)
        if rangesAreText: #then output field type must be cast to integer
            fmString = fieldMappings.exportToString()
            newfmString = ""
            for s in fmString.split(";"):
                for f in [lfromField,ltoField,rfromField,rtoField]:
                    if f in s:
```

Again, the text following the number sign is not run, but the code preceding it on the same line is executed. Comments are used to provide generic descriptions of who authored the tool and how it works, as well as any pertinent details about specific code elements. Commenting helps others understand how the script works, but it can also help the original coder remember how the code was created and why.

Commenting is not required for a script to work properly. However, using comments is a good coding practice—both as a service to others who use the script as well as a reminder to yourself of how the code works. At a minimum, each script should contain a heading section that describes what the script does, who created it and when, and what the requirements are for the script to run.

> **Note:** *Using the number sign (#) is not the only way to indicate a line is a comment. Sometimes double number signs (##) are used instead, although this is often reserved for temporarily commenting out code, or temporarily turning code into comments so as not to lose it. The effect is the same: Code lines that start with number signs (# or ##) are not run.*

In addition to comments, you can use blank lines to organize code. Technically, blank lines in a Python script are ignored on execution but they make it easier to read the code. Usually, blank lines are used to keep lines of related code together and separate from other sections. Like comments, blank lines are not required for a script to work properly, but they make code easier to read.

# 4.20 Working with code in the PythonWin editor

Because you will be working a lot with scripts in an editor like PythonWin, a few observations are in order to work with code in scripts. Consider the earlier example of the script with a `while` loop, as shown in the figure. ➔

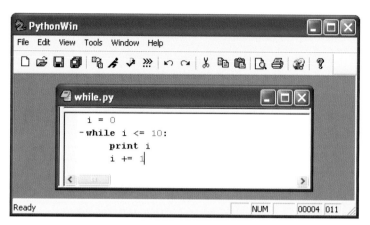

First, notice that the location of the cursor is shown in the lower-right corner of the PythonWin interface, on the status bar. In the while.py script, the cursor is placed at the end of the script. The first number on the status bar is the line number (4) and the second number is the character position (11). For longer scripts, it will be useful to show the line numbers directly within the script window itself. You can change the display options by clicking View > Options from the PythonWin menu bar. On the PythonWin Options dialog box, click the Editor tab. On the Editor tab under Margin Widths, increase the value of Line Numbers—for example, to 30. ➔

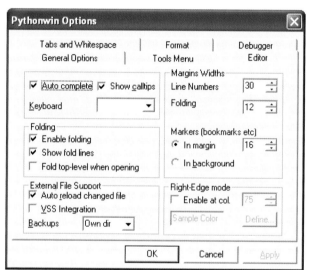

This makes the line numbers visible, as shown in the figure below.

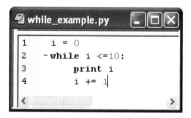

Second, notice the minus sign (-) in front of the line of code starting with `while` in the graphic above. Clicking the minus sign collapses the block of code.

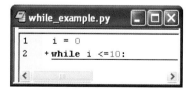

This can be useful to for increasing the readability of longer scripts that have many blocks of code. Clicking the plus sign (+) expands the block of code again.

Third, by default, spaces and tabs do not show up in code. To make them visible, as shown in the figure, click View > Whitespace from the PythonWin menu bar.

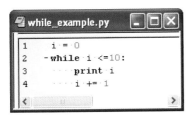

This allows you to see the number of spaces used in the block of code. This can be useful for ensuring consistent indentation.

## 4.21 Following coding guidelines

Python enforces certain coding standards, and code that does not meet these standards will produce errors. In addition, there are coding guidelines that can assist in making sure your code is not only error free, but also efficient, portable, and easy to follow. These guidelines are formalized in the *Style Guide for Python Code*, also known as PEP 8. This is part of a larger set of Python Enhancement Proposals (PEPs).

Following are a number of coding guidelines for the topics covered so far. These guidelines reflect PEP 8 as well as other considerations. As you learn more about Python and start writing your own scripts, it is a good idea to become familiar with the complete style guide, at http://www.python.org/dev/peps/pep-0008. Some pointers follow.

*Note: It is common to see variable names that consist of several words, in which the first letter of each word is capitalized, except for the first word—for example,* `myIntegerVariable`. *This capitalization does not follow the Style Guide for Python Code and is not recommended.*

### Variable names

- Start with a character and avoid using special characters such as an asterisk (*).

- Use all lowercase, such as `mycount`.

- Underscores (_) can be used if they improve readability, such as `count_final`.

- Use descriptive variable names and avoid using slang terms or abbreviations.

- Keep variable names short.

**Script names**

- Script names should follow the preceding variable naming guidelines—that is, use all lowercase, and underscores can be used to improve readability.

**Indentation**

- The use of four spaces is recommended to define each indentation level.

- Never mix tabs and spaces.

**Comments**

- Scripts should contain adequate commenting. Each logical section should have an explanation.

- Each script tool or function should have a header that contains the script name, a description of how the script works, its requirements, who wrote it, and when it was written.

It is important to recognize that following the Python guidelines is not required—that is, if you break from the guidelines, it will not necessarily result in a syntax error. However, following the guidelines will improve the consistency and readability of your code.

*Note: Although Python script names may have the first letter of each word capitalized, it does not follow the Style Guide for Python Code and is not recommended.*

# Points to remember

This chapter introduces you to the fundamentals of the Python language. There is obviously much more to learn about Python functionality, but these fundamentals will allow you to get started writing geoprocessing scripts.

A few points to remember about Python:

- Python code can be evaluated directly in the interactive Python interpreter, but to save your code you have to create a Python script file, which is saved with a .py file extension.

- Python uses expressions, which represent a value or values, and statements, which do something.

- Variable names should be all lowercase and contain only characters, digits, and the underscore (_). Variables are given a value using an assignment statement.

- Python contains a number of standard built-in functions, which can be used by employing a function call. To use a function that is not one of the built-in functions, you first need to import the module, and then call the function using `<module>.<function>`.

- Strings can be manipulated using Python's built-in methods, including finding specific substrings, joining strings together, splitting strings based on a separator, stripping characters from the start and end of a string, converting the case of a string, and more.

- Lists are a versatile Python data type. Specific elements from a list can be obtained using the element's index number, starting with zero (0)—for example, `mylist[0]`. There are a number of built-in Python functions and methods to manipulate lists. These include ways to sort a list, slice a list into smaller lists, delete elements, append elements, insert elements, and more.

- Workflow in Python scripts can be controlled using branching and loop structures. These structures use indentation to identify blocks of code. Indentation is part of the Python syntax and is not optional.

- Python syntax is case sensitive, for the most part.

# Part 2
**Writing scripts**

# Chapter 5
## Geoprocessing using Python

## 5.1 Introduction

This chapter describes the ArcPy site package, which allows for a close integration of ArcGIS and Python. ArcPy modules, classes, and functions, which give access to all the geoprocessing tools in ArcGIS, are introduced. Several additional nontool functions related to running geoprocessing tasks are also covered, including establishing environment settings, setting paths to data, and managing licenses.

## 5.2 Using the ArcPy site package

The geoprocessing functionality of ArcGIS can be accessed through Python using the ArcPy site package. A site package in Python is like a library of functions that add functionality to Python. The site package works very much like a module, but a package contains multiple modules as well as functions and classes.

ArcPy was introduced in version 10 of ArcGIS with the goal of making Python scripting easier and more powerful. Prior to ArcGIS 10, the geoprocessing functionality of ArcGIS was accessed through Python using the `ArcGISscripting` module. Scripts written for earlier versions of ArcGIS typically use this module. The focus in this book is on ArcPy, and the older module is not covered in detail. However, because you may sometimes be working with older scripts, the `ArcGISscripting` module is briefly explained in section 5.4. This module is still supported in ArcGIS 10, so older scripts using this module continue to work.

ArcPy is organized in modules, functions, tools, and classes, which are described later in this chapter.

# 5.3 **Importing ArcPy**

Working with ArcPy starts with importing the site package. A typical geo-processing script therefore starts with the following line of code:

```
import arcpy
```

Once you import ArcPy, you can run all the geoprocessing tools found in the standard toolboxes installed with ArcGIS.

ArcPy contains many modules, including two specialized ones: a map automation module (`arcpy.mapping`) and a map algebra module (`arcpy.sa`). To import these modules, you can use the following syntax:

```
import arcpy.mapping
```

Once you import ArcPy or one of its specialized modules, you can start using its modules, functions, and classes.

One of the first tasks typically is to set the current workspace. For example, here is how you would set the current workspace to C:\Data:

```
import arcpy
arcpy.env.workspace = "C:/Data"
```

*Note: Remember that you should not use a backslash (\) for paths because Python views it as an escape character.*

Notice that the path is a string variable.

Environment settings are exposed as properties of the ArcPy `env` class. Classes are explained in more detail in section 5.8. These properties can be used to retrieve the current values or to set them. In the preceding code, `env` is a class and `workspace` is a property of this class. When using classes, the syntax is as follows:

```
arcpy.<class>.<property>
```

You will see more examples of this syntax later in this chapter.

Often you may not need to use the entire module. You can use the `from-import` statement to import only a portion of a module. The following code imports just the `env` class. Instead of accessing environments using `arcpy.env`, you can simplify it to `env`.

```
from arcpy import env
env.workspace = "C:/Data"
```

You can further control the importing of modules by giving a module or part of a module a custom name using the `from-import-as` statement as follows:

```
from arcpy import env as myenv
myenv.workspace = "C:/Data"
```

Although using custom names does not shorten the length of your code, it can make it easier to read.

The `from-import-as-*` statement goes a step further. In this case, the contents of the module are imported directly into the namespace, meaning that you no longer need to add a prefix to the contents, such as "myenv," or use the module name.

```
from arcpy import env as *
workspace = "C:/Data"
```

*Note: The use of the* `from-import-as-*` *statement can reduce the length of your code, but you have to be very careful because other objects, variables, and modules that have the same name will be overwritten. In general, it is recommended that you stick with* `from arcpy import env`.

## 5.4  Working with earlier versions of ArcGIS

The ArcPy site package was introduced with ArcGIS 10. Prior to the introduction of ArcPy, geoprocessing tools were accessed from Python using the `ArcGISscripting` module. Thus, Python scripts created prior to ArcPy use a syntax that is slightly different from the `import arcpy` statement to import geoprocessing functionality.

Using the 9.3 version of the geoprocessor object, the syntax is as follows:

```
import ArcGISscripting
gp = ArcGISscripting.create(9.3)
```

Using the pre-9.3 version of the geoprocessor object, the syntax is as follows:

```
import ArcGISscripting
gp = ArcGISscripting.create()
```

Notice that, in both cases, first the `ArcGISscripting` module is imported, and then a geoprocessor object is created. All geoprocessing capabilities of ArcGIS are exposed as methods of the geoprocessor object. In ArcPy, it is no longer necessary to create a geoprocessor object.

Prior to the introduction of the `ArcGISscripting` module, Python scripts could access the geoprocessing tools using the Python `win32com` module to create the geoprocessor object. Here is what the code looks like using the `win32com` module:

```
import win32com.client
gp = win32com.client.Dispatch("esriGeoprocessing.GpDispatch.1")
```

Once the geoprocessor object is created using one of these three methods, the rest of the syntax is relatively similar, although not identical, to the use of ArcPy. For example, the code to set the current workspace is

```
gp.workspace = "C:/Data"
```

The Python `win32com` module is no longer installed with ArcGIS 10. As a result, any scripts that use the `win32com` module do not work under the default installation. However, the installation of the PythonWin editor includes an installation of the `win32com` module, which allows scripts using the module to work.

Despite the similarity, it can be a bit confusing to work with these earlier versions of ArcGIS and ArcPy at the same time. In this book, the focus is on the use of ArcPy because it has many advantages over the earlier versions. Many existing Python scripts (including many written by Esri) were written using the `ArcGISscripting` module. These scripts continue to work because ArcGIS 10 continues to support them. However, much of the additional functionality introduced in ArcGIS 10 is not available in the earlier versions of ArcGIS.

If you are just getting started with Python scripting, you will most likely need to learn how to work with the ArcPy site package only. If you need to edit older scripts, you may need to learn a bit more about the `ArcGISscripting` and `win32com` modules. Please note, however, that ArcGIS Desktop Help for version 10 no longer includes a description of these modules or any sample code using these modules. As a result, you would need to fall back on the Help for version 9.3. These Help files remain available online at http://webhelp.esri.com/ArcGISdesktop/9.3.

*__Note:__ The book includes no further details on the `ArcGISscripting` or `win32com` modules.*

## 5.5 **Using tools**

ArcPy gives you access to ArcGIS for Desktop geoprocessing tools. When working with geoprocessing tools in Python, the tools are referred to by name. This does not correspond exactly to the tool label, which is how the tool appears in ArcToolbox. A tool name is generally very similar to the tool label but contains no spaces. For example, the name of the Add Field tool in the Data Management toolbox is AddField.

In addition to using the tool name rather than the tool label, a reference to a particular tool also requires the toolbox alias. It is because multiple tools in different toolboxes can share the same name. For example, there are several Clip tools: one in the Analysis toolbox and one in the Data Management toolbox. The toolbox alias is not the same as either the name or the label of the toolbox—it is typically an abbreviated version. For example, the alias of the Data Management toolbox is "management."

The Clip tool in the Data Management toolbox is therefore referenced as Clip_management. Notice that the name of the toolset (Raster > Raster Processing) is not referenced in any way.

There are two ways to access a tool in a line of Python code. The easiest way to call a tool is to call its corresponding function. All tools are available as functions in ArcPy. An ArcPy function is a defined bit of functionality that does a specific task. The syntax for calling a tool by its function is as follows:

```
arcpy.<toolname_toolboxalias>(<parameters>)
```

For example, the following code runs the Clip tool:

```
import arcpy
arcpy.env.workspace = "C:/Data"
arcpy.Clip_analysis("streams.shp", "study.shp", "result.shp")
```

Tools are also available in modules that match the toolbox alias name. An alternative way to access a tool is to first call the toolbox as a module and then the tool as a function, followed by the tool's parameters. The syntax is as follows:

```
arcpy.<toolboxalias>.<toolname>(<parameters>)
```

Here is what the example looks like for running the Clip tool:

```
import arcpy
arcpy.env.workspace = "C:/Data"
arcpy.analysis.Clip("streams.shp", "study.shp", "result.shp")
```

Both methods are correct, and the approach you use is a matter of preference and coding habits.

Just a few quick reminders regarding Python syntax:

- Python is case sensitive, so `Clip` is correct, but `clip` is not.

- The use of spaces, or whitespace, in a line of code has no effect on its execution. For example, `workspace="C:/Data"` is the same as `workspace = "C:/Data"`. Whitespace is recommended to improve readability but is not required. However, do not include spaces between modules, functions, classes, methods, and properties, so `env.workspace` is correct, but `env. workspace` is not. Also do not include spaces between functions and their arguments, so

*Note: The coding style adopted for this book uses the* `arcpy.Clip_analysis` *style of calling tools, but you may see the other style in example scripts from other sources.*

<toolname>(<parameters>) is correct, but
<toolname> (<parameters>) is not.

- Quotation marks in Python are straight and not curly, so use " " and not “ ”. When typing code using a Python editor, the correct style of quotation marks is entered automatically. However, incorrect quotation marks can occur when copying code from other applications, such as a Microsoft Word document or a PDF file.

A key aspect of running geoprocessing tools is to get the syntax right for the parameters. Every geoprocessing tool has parameters, required and optional, that provide the tool with the information it needs for execution. Common parameters are input datasets, output datasets, and keywords that control the execution of the tool. Parameters themselves have properties such as the following:

- Name: a unique name for each tool parameter
- Type: the type of data expected, such as feature class, integer, string, or raster
- Direction: whether the parameter defines input or output values
- Required: whether a value must be provided for a parameter or is optional

The documentation of each tool describes its parameters and properties. Once a valid set of parameters is provided, the tool is ready to be run. Most parameters are specified as a simple string. Strings consist of text that identifies a parameter value, such as a path to a dataset or a keyword.

The Clip tool's documentation, as shown in the figure, describes its parameters.

| **Parameter** | **Explanation** | **Data Type** |
|---|---|---|
| in_features | The features to be clipped. | Feature Layer |
| clip_features | The features used to clip the input features. | Feature Layer |
| out_feature_class | The feature class to be created. | Feature Class |
| cluster_tolerance (Optional) | The minimum distance separating all feature coordinates as well as the distance a coordinate can move in X or Y (or both). Set the value to be higher for data with less coordinate accuracy and lower for data with extremely high accuracy. | Linear unit |

The Clip tool has four parameters, with the last one (cluster_tolerance) being optional. The syntax of the Clip tool is

```
Clip_analysis(in_features, clip_features, out_feature_class, {cluster_tolerance})
```

The name of the Clip tool is followed by the tool's parameters in paren-
theses. Parameters are separated by a comma (,). Optional parameters are
surrounded by curly brackets ({ }).

The syntax of geoprocessing tools typically follows the same general pat-
tern, as follows:

- Required parameters come first, followed by optional parameters.

- The input datasets are usually the first parameter or parameters,
  followed by the output dataset if there is one. Next are additional
  required parameters, and finally, optional parameters.

- Parameter names for input datasets are prefixed by "in_" (such as,
  in_data, in_features, in_table, in_workspace) and parameter names
  for output datasets are prefixed by "out_" (such as, out_data, out_fea-
  tures, out_table).

Listing required parameters first makes it easy to simply leave out the
optional parameters when they are not needed. Sometimes, however, some
of the optional parameters need to be set. Because parameters need to be
specified in the order that they are listed in the tool syntax, it can mean that
some optional parameters may need to be skipped.

Consider, for example, the syntax of the Buffer tool:

```
Buffer_analysis (in_features, out_feature_class, buffer_distance_or_ ➜
➜ field, {line_side}, {line_end_type}, {dissolve_option}, {dissolve_field})
```

A code example of the Buffer tool is as follows:

```
import arcpy
arcpy.env.workspace = "C:/Data/study.gdb"
arcpy.Buffer_analysis("roads", "buffer", "100 METERS")
```

Using this example, how would you specify the optional dissolve_option
parameter and skip the other optional parameters that follow the required
parameters? It can be accomplished in different ways, as follows:

- By setting the optional parameters using an empty string ("") or the
  number sign ("#")

- By specifying by name the parameter that needs to be set, bypassing
  all others

The Buffer tool has three required parameters and four optional parameters. To specify a dissolve option and the field to use in this dissolve, two optional parameters need to be skipped. This can be done in three ways:

```
arcpy.Buffer_analysis("roads", "buffer", "100 METERS", "", "", "LIST", ➡
➡ "Code")

arcpy.Buffer_analysis("roads", "buffer", "100 METERS", "#", "#", "LIST", ➡
➡ "Code")

arcpy.Buffer_analysis("roads", "buffer", "100 METERS", dissolve_ ➡
➡ option="LIST", dissolve_field="Code")
```

In each of these three cases, the optional parameters line_side and line_end_type are left as default values.

In the examples so far, the parameters of the tool use the actual file name (for example, "roads"). This means the file names are *hard-coded*. That is, the parameters are not set as variables, but use the values directly. Although this syntax is correct and works fine, it is often more useful to make your code flexible by using variables for parameters instead of using file names. First, create a variable and assign it a value. Then you can use the variable for a parameter. These variable values are passed to the tool. For example, in the case of the Clip tool, it would look as follows:

*Note: Although each of these three options is correct, the examples in this book typically use the empty string ("").*

```
import arcpy
arcpy.env.workspace = "C:/Data"
infc = "streams.shp"
clipfc = "study.shp"
outfc = "result.shp"
arcpy.Clip_analysis(infc, clipfc, outfc)
```

In this example script, the names of the datasets are still hard-coded in the script itself but not in the specific line of code that calls the Clip tool. The next logical step is to have the values of the variables provided by a user or another tool, which means the actual file names would no longer appear in the script. For example, the following code runs the Copy tool, and the input and output feature classes are obtained from user input using the GetParameterAsText function:

```
import arcpy
infc = arcpy.GetParameterAsText(0)
outfc = arcpy.GetParameterAsText(1)
arcpy.Copy_management(infc, outfc)
```

Functions are covered later in this chapter and the GetParameterAsText and related functions are covered in chapter 13. Setting tool parameters based on user input is commonly used for script tools. Working with variables in this way gives you more flexibility and makes much of your code reusable.

Here are just a few quick reminders regarding variable names in Python. Variable names can consist of any combination of valid characters. However, adopting a consistent coding style is recommended. For example, the *Style Guide for Python Code* recommends using all lowercase characters and using underscores (_) only if it improves readability—for example, `my_clip` or `clip_result`. Variable names should also be kept short (but meaningful) to limit the need for typing and associated typos—for example, `clipfc` instead of `clipfeatureclass`.

ArcPy returns the output of a tool as a result object. When the output of a tool is a new or updated feature class, the result object includes the path to the dataset. For other tools, however, the result object can consist of a string, a number, or a Boolean value. One of the advantages of result objects is that you can keep track of information about the running of tools. This includes not only the output, but also messages and parameters.

For example, in the following code, a geoprocessing tool is run and the output is returned as a result object:

```
import arcpy
arcpy.env.workspace = "C:/Data"
mycount = arcpy.GetCount_management("streams.shp")
print mycount
```

This code displays the string representation of the result object. For example:

```
3153
```

When the output of a tool consists of a feature class, the result object includes the path to the dataset. For example, the following code runs the Clip tool:

```
import arcpy
arcpy.env.workspace = "C:/Data"
myresult = arcpy.analysis.Clip("streams.shp", "study.shp", "result.shp")
print myresult
```

Running the code displays the string representation of the path to the output dataset:

```
C:/Data/result.shp
```

The result object has properties and methods, which are covered in more detail later in this chapter.

The result object can be used as an input to another function. For example, in the following code, a feature class is buffered using the Buffer tool.

The output polygon feature class is returned as an object and the object is used as the input to the Get Count tool, as follows:

```
import arcpy
arcpy.env.workspace = "C:/Data/study.gdb"
buffer = arcpy.Buffer_analysis("str", "str_buf", "100 METERS")
count = arcpy.GetCount_management(buffer)
print count
```

Although many tools have only a single output, some tools have multiple outputs. The `getOutput` method of the result object can be used to obtain a specific output by using an index number. A more generic way to get a geoprocessing result follows:

```
import arcpy
arcpy.env.workspace = "C:/Data/study.gdb"
buffer = arcpy.Buffer_analysis("str", "str_buf", "100 METERS")
count = arcpy.GetCount_management(buffer).getOutput(0)
print str(count)
```

You can probably see where this is going. You can create a series of geo-processing operations, just as in ModelBuilder, and only the final desired output is returned back to the application that called the script.

## 5.6 Working with toolboxes

When the ArcPy site package is imported into Python, all the system tool-boxes are available. When custom tools are created and stored in a custom toolbox, these tools can be accessed in Python only by importing the cus-tom toolbox. So even if a custom toolbox has been added to ArcToolbox in ArcMap or ArcCatalog, Python is not aware of this toolbox until it has been imported. This is accomplished using the `ImportToolbox` function. The following code illustrates how to import a toolbox:

```
import arcpy
arcpy.ImportToolbox("C:/Data/sampletools.tbx")
```

Notice that the `ImportToolbox` function references the actual file on disk—that is, the toolbox (.tbx) file, not the name of the toolbox.

After importing a toolbox, the syntax for accessing a tool in Python is as follows:

```
arcpy.<toolname>_<toolboxalias>
```

This syntax is exactly the same as the syntax for accessing the tools from the system toolboxes. ArcPy depends on toolbox aliases to access and run the correct tool. System toolboxes have well-defined aliases, but this may not be the case for custom toolboxes. The alias of a toolbox is different from both the name (which is the file name of the .tbx file) and the label (which is the display name of the toolbox) and needs to be specified separately; there is no default alias. It is therefore good practice to always define a custom toolbox alias. However, if a toolbox alias is not defined, a temporary alias can be set as the second parameter of the `ImportToolbox` function, as follows.

```
arcpy.ImportToolbox("sampletools.tbx", mytools)
```

Once the alias is set, tools in the toolbox can be accessed using Python. For example, if the `sampletools.tbx` file contains a tool called MyModel, the syntax to access this tool would look as follows:

```
arcpy.MyModel_mytools(<parameters>)
```

Or alternatively:

```
arcpy.mytools.MyModel(<parameters>)
```

The `ImportToolbox` function can also be used to add geoprocessing services from an Internet or local server.

Although ArcPy provides access to all the geoprocessing tools in ArcGIS, the tools that are available depend on the product (ArcGIS for Desktop Basic, ArcGIS for Desktop Standard, and ArcGIS for Desktop Advanced) and whether any extensions are installed and licensed. In addition, custom toolboxes may be installed, which adds new tools.

Once a particular tool is identified, the tool's syntax can be accessed from Python using the `Usage` function. For example, the following code prints the syntax of all the tools in the Editing Tools toolbox:

```
import arcpy
tools = arcpy.ListTools("*_analysis")
for tool in tools:
print arcpy.Usage(tool)
```

The syntax can also be accessed in the Help file for each tool, but the `Usage` function gives you access to the syntax within Python.

Another way to access the syntax directly is to use Python's built-in `help` function. For example, the following code prints the syntax of the Clip tool:

```
print help(arcpy.Clip_analysis)
```

# 5.7 **Using functions**

A function in Python is a specific bit of functionality that does a specific task. All geoprocessing tools are provided as functions. In addition, ArcPy provides a number of functions that are not tools. Functions can be used to list datasets, retrieve properties of a dataset, check for the existence of data, validate names of datasets, and perform many other useful tasks. These functions are designed for Python workflows and therefore are available only from ArcPy and not as tools in ArcToolbox. So in ArcPy, all tools are functions, but not all functions are tools.

The general form of a function is very similar to that of a tool. A function has parameters (also referred to as arguments), which can be required or optional. A function returns values. Returned values for most functions are result objects. It can be the path to a dataset, a string, a number, a Boolean value, or a geoprocessing object.

The syntax of a function is the same as for tools:

```
arcpy.<functionname>(<arguments>)
```

For example, the following code determines whether a particular dataset exists, and then prints either `True` or `False`:

```
import arcpy
print arcpy.Exists("C:/Data.streams.shp")
```

The `arcpy.Exists` function returns a Boolean value. Other functions return other types of values, including strings and numbers.

There are a large number of functions, which can be divided into the following categories:

- Cursors
- Describing data
- Environment and settings
- Fields
- Geodatabase administration
- General
- General data functions
- Getting and setting parameters
- Licensing and installation
- Listing data
- Log history
- Messaging and error handling
- Progress dialog boxes
- Raster
- Spatial reference and transformations
- Tools and toolboxes

These categories are created primarily to provide a logical organization of the functions, but the names of these categories do not appear in Python syntax. So unlike tools, these functions are always called directly, without reference to the categories. ArcGIS Desktop Help provides a complete list of ArcPy functions and a detailed description of each. Several of these functions are revisited in later chapters of the book.

Technically speaking, all geoprocessing tools are functions in ArcPy, and they are accessed like any other Python function. To avoid confusion, ArcPy functions are divided into tool functions and nontool functions. There are a number of important distinctions between the two:

- The documentation is located in different sections of ArcGIS Desktop Help. Tools are documented in the Geoprocessing Tool Reference. It can also be obtained from the tool dialog box when running a tool from ArcToolbox. Nontool functions are documented only in the ArcPy documentation.

- Tools are licensed by product level (ArcGIS for Desktop Basic, ArcGIS for Desktop Standard, and ArcGIS for Desktop Advanced) and by extension (ArcGIS 3D Analyst, ArcGIS Network Analyst, ArcGIS Spatial Analyst, and more). The Geoprocessing Tool Reference indicates the license level that is required for each tool. Nontool functions, on the other hand, are not licensed separately. All the ArcPy nontool functions are installed with ArcPy, independent of the license level.

- Tools produce geoprocessing messages, which can be accessed through a variety of functions. Nontool functions do not produce these messages.

- Calling tools requires the use of the toolbox alias or the module name, whereas nontool functions do not.

- Tools return a result object, whereas nontool functions do not.

## 5.8  Using classes

Many tool parameters are straightforward—for example, feature classes, field names, and numerical values. They are relatively easy to work with and can often be specified using a simple string value. Some tool parameters, however, are more complex—for example, using a coordinate system as a parameter, or a reclassification table when working with raster data. This is where classes come into play. ArcPy classes are often used as shortcuts for tool parameters that would otherwise have a more complicated equivalent. Classes can be used to create objects, and once the object is created, its properties and methods can be used. So, the spatial reference or the reclassification table is passed as an object.

Consider the example of a class property. Earlier in this chapter, you worked with the ArcPy env class. Environment settings are exposed as properties of the env class. For example, workspace is a property of the env class, so the syntax becomes env.workspace.

The syntax for setting the property of a class is

```
<classname>.<property> = <value>
```

As discussed earlier, the code to set the current workspace is as follows:

```
import arcpy
arcpy.env.workspace = "C:/Data"
```

Another frequently used ArcPy class is the SpatialReference class. This class has a number of properties, including the coordinate system parameters. To work with these properties, however, the class first has to be *instantiated*. The syntax for using a method to initialize a new instance of a class is

```
arcpy.<classname>(parameters)
```

The code to initialize a new instance of the SpatialReference class is as follows:

```
import arcpy
prjfile = "C:/Data/myprojection.prj"
spatialref = arcpy.SpatialReference(prjfile)
```

In this example, SpatialReference is the class that creates the spatialref object by reading an existing projection (.prj) file. The actual .prj file already exists on disk and is used to create the object. Once the object is created, you can work with the properties of the object. For example, in the case of the SpatialReference class, you can work with any of the specific parameters that define a spatial reference file, such as the coordinate system parameters, tolerances, and domains.

For example, the following code creates a spatial reference object based on an existing .prj file, and then uses the name property to get the name of the spatial reference:

```
import arcpy
prjfile = "C:/Data/streams.prj"
spatialref = arcpy.SpatialReference(prjfile)
myref = spatialRef.name
print myref
```

Running the code prints the name of the spatial reference, something like
the following:

```
NAD_1983_StatePlane_Florida_East_FIPS_0901_Feet
```

Classes are often used to avoid having to use long and complicated strings.
A good example is using classes for complex tool parameters. Most tool
parameters are defined using simple strings, including dataset names, paths,
keywords, field names, domain names, tolerances, and others. However,
some parameters are harder to define by simple strings because the parameters require more properties. For example, think of the coordinate system
of a feature class. For a shapefile, it is stored in a .prj file—a regular text file
with the file extension .prj. Opening a .prj file in Notepad looks something
like the example in the figure.

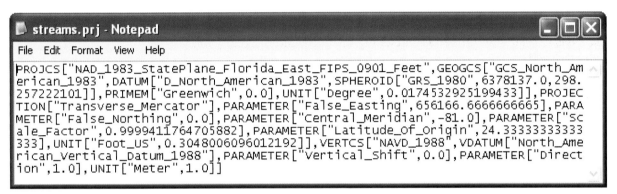

Working with this type of string can be a bit cumbersome. It would be
much easier if you could refer to it simply by using the name of the coordinate system or by referencing the .prj file that contains the string value.
Working with the `SpatialReference` class is the way to do it.

For example, a `SpatialReference` object can be created and used to
define the output coordinate system of a new feature class. The new feature class is created using the Create Feature Class tool. The syntax of the
Create Feature Class tool is as follows:

```
CreateFeatureclass_management(out_path, out_name, {geometry_type}, →
    → {template}, {has_m}, {has_z}, {spatial_reference}, {config_keyword}, →
    → {spatial_grid_1}, {spatial_grid_2}, {spatial_grid_3})
```

The following code creates a spatial reference object and uses it to define
the output coordinate system of a new feature class:

```
import arcpy
out_path = "C:/Data"
out_name = "lines.shp"
prjfile = "C:/Data/streams.prj"
spatialref = arcpy.SpatialReference(prjfile)
arcpy.CreateFeatureclass_management(out_path, out_name, "POLYLINE", "", ➡
➡ "", "", spatialref)
```

Using the `SpatialReference` object is a lot easier than trying to work
with the actual string value contained in the .prj file.

> *Note: As a reminder, optional tool parameters can be left out of the syntax unless they
> are followed by an optional parameter that is being specified differently from the default.
> A blank string ("") signifies the use of default values for optional parameters. In the
> preceding example code, the spatial reference parameter is preceded by three optional
> parameters—hence the use of three empty strings.*

# 5.9 **Using environment settings**

Environment settings are essentially hidden parameters that influence how a
tool runs. You have already seen how to set the environments in Python using
the `env` class. This section covers these settings in a bit more detail, because
environments are fundamental to controlling geoprocessing workflows.

Environment settings are exposed as properties of the `env` class. These
properties can be used to retrieve the current values or to set them. Each
property has a name and a label. The labels are displayed on the Environment Settings dialog box in ArcGIS, but Python works with names only.
The syntax for accessing the properties from the environment class is

```
arcpy.env.<environmentName>
```

For example, to set the current workspace, the following code is used:

```
import arcpy
arcpy.env.workspace = "C:/Data"
```

Alternatively, environments can be accessed using the `from-import`
statement:

```
import arcpy
from arcpy import env
env.workspace = "C:/Data"
```

The env class also has many other properties. A complete list can be found in the ArcPy documentation. Some important properties include the extent, the output coordinate system, the scratch workspace, and the XY domain. Some properties are specific either to feature classes or to raster datasets. For example, cell size, compression, and mask are used for raster datasets only. The following code sets the cell size to 30:

```
import arcpy
from arcpy import env
env.cellSize = 30
```

The properties of the env class not only specify environments, but can also be used to retrieve their current values. For example, the following code retrieves the current settings for the XY tolerance:

```
import arcpy
from arcpy import env
print env.XYTolerance
```

Running the code prints the current value of the XY Tolerance parameter. The default value of None is printed unless the value has previously been set.

To get a complete list of properties, you can use the ArcPy ListEnvironments function:

```
import arcpy
print arcpy.ListEnvironments()
```

Running the code prints the alphabetical list of all properties.

There is one additional environment setting that is of special interest. You may recall from chapter 2, the geoprocessing options include the option to overwrite the outputs of geoprocessing operations. In ArcMap, this is not part of the Environment Settings dialog box but is a separate option on the menu bar under Geoprocessing > Options. In Python, this is a property of the env class. The default value of this overwriteOutput property is False. The following code sets the value to True:

```
import arcpy
from arcpy import env
env.overwriteOutput = True
```

Environment settings are revisited again in later chapters, especially how to transfer settings from one script to another.

# 5.10 **Working with tool messages**

When tools are run, messages are written about the tool's success or failure of execution. Communication between tools and users is accomplished using messages. Typical information in these messages consists of the following:

- The exact time when running the tool started and ended

- The parameter values used to run the tool

- Information about the progress in running the tool (information messages)

- Warnings of potential problems in running the tool (warning messages)

- Any errors that prevented the tool from running (error messages)

When a tool is run from ArcToolbox, these geoprocessing messages appear on the progress dialog box when background processing is disabled. When a tool is run from the Python window, only error messages that indicate a particular situation prevented the tool from running appear. When a tool is run from within an ArcGIS for Desktop application, messages also appear in the Results window. A typical Results window after a tool has been run looks like the example in the figure. ➜

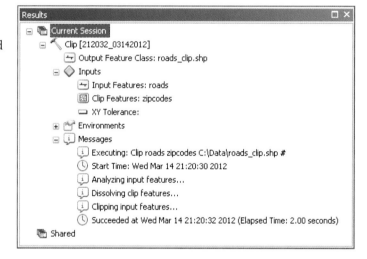

When a stand-alone Python script is run, these messages are not added to the Results window. Instead, you can obtain them from within the script. You can print the messages, write them to a file, or query them.

All messages have a severity property. This property is an integer with a value of 0 (information), 1 (warning), or 2 (error). Table 5.1 describes these three levels of severity in more detail.

**Table 5.1  Severity of messages**

| Severity | Description |
| --- | --- |
| Severity = 0 Information message | Information messages provide information about tool execution. This includes general information such as the tool's progress, the start and completion time of the tool, and information about the tool results. |
| Severity = 1 Warning message | Warning messages indicate a possible problem. This could be a situation that may cause a problem during the tool's execution or a situation where the result may not be what you might expect. Warning messages do not prevent a tool from being executed, but they do warrant inspection. |
| Severity = 2 Error message | Error messages indicate that the tool has failed. Typically, this means that one or more parameter settings are invalid. |

Both warning and error messages are accompanied by a six-digit ID code. The ID codes are documented, and the description of each ID code may be helpful in identifying the causes of potential problems and how they can be dealt with.

Messages from the last tool run are maintained by ArcPy and can be retrieved using the `GetMessages` function. This function returns a single string containing all the messages from the tool that was last run. The messages can be filtered by providing the severity argument.

The basic syntax for retrieving messages and printing them is

```
print arcpy.GetMesssages()
```

For example, when a Clip tool is run, the messages can be retrieved as follows:

```
import arcpy
arcpy.env.workspace = "C:/Data"
infc = "streams.shp"
clipfc = "study.shp"
outfc = "result.shp"
arcpy.Clip_analysis(infc, clipfc, outfc)
print arcpy.GetMessages()
```

Running the code results in a list of messages similar to the following:

```
Executing: Clip C:\Data\streams.shp C:\Data\study.shp C:\Data\result.shp #
Reading Features...
Cracking Features...
Assembling Features...
Succeeded at Fri Apr 30 17:12:05 2010 (Elapsed Time: 2.00 seconds)
```

Individual messages can be retrieved using the `GetMessage` function (note, this is different from the `GetMessages` function). This function has only one parameter, which is the index position of the message. For example, the following code retrieves the first message only:

```
print arcpy.GetMessage(0)
```

The result is

```
Executing: Clip C:\Data\streams.shp C:\Data\study.shp C:\Data\result.shp #
```

**Note:** *The index position starts at zero (0).*

The number of messages from the last tool that is run can be obtained using the `GetMessageCount` function. This function is particularly useful for retrieving just the last message. Since you typically will not know in advance how many messages may have resulted from running a tool, you can use the message count to retrieve the last message. The code to obtain the message count is

```
arcpy.GetMessageCount()
```

To retrieve the last message only, you would use the following:

```
count = arcpy.GetMessageCount()
print arcpy.GetMessage(count-1)
```

The result would be

```
Succeeded at Fri Apr 30 17:12:05 2010 (Elapsed Time: 2.00 seconds)
```

In addition to getting the number of messages, you can also query the maximum severity of the messages using the `GetMaxSeverity` function as follows:

```
print arcpy.GetMaxSeverity()
```

In the prior example of running the Clip tool, running the code would return the value of 0 because there are only information messages.

Although the `GetMessage`, `GetMessageCount`, and `GetMaxSeverity` functions are useful, in practice the `GetMessage` function is the most widely used. Messages are most important when a tool fails and so the `GetMessage` function is commonly used in combination with an error-handling technique. This is covered in chapter 11.

The functions discussed thus far allow you to retrieve the messages from the last tool that was run because the messages are maintained by ArcPy. However, as soon as another tool is run, you can no longer retrieve messages from tools run prior to that. To retrieve messages even after multiple tools have been run, the result class can be used to create a result object. The result object can then be used to retrieve and interpret geoprocessing tool messages. So rather than a tool being run for output files, the result of the geoprocessing operation is returned as an object. For example:

```
import arcpy
arcpy.env.workspace = "C:/Data"
result = arcpy.GetCount_management("streams.shp")
```

The result class has a number of properties and methods. The `messageCount` property returns the number of messages and `getMessage` returns a specific message. For example, running the following code retrieves the number of messages, followed by the last message:

```
import arcpy
arcpy.env.workspace = "C:/Data"
result = arcpy.GetCount_management("streams.shp")
count = result.messageCount
print result.getMessage(count-1)
```

Notice that the syntax is similar, but not identical, to using the general message function. When using `arcpy.GetMessage()`, you are calling a function, whereas using `<objectname>.getMessage()`, you are retrieving the properties of an object. The result class has a number of advantages over calling message functions, most notably the fact that messages can be maintained after running multiple tools. The result class also has a number of additional properties and methods, including options to count the number of outputs and the ability to work with specific outputs from a geoprocessing tool.

# 5.11 **Working with licenses**

Running geoprocessing tools requires a license for an ArcGIS product, such as ArcGIS for Desktop or ArcGIS for Server. This is true for running a tool from ArcToolbox within any ArcGIS for Desktop application, but it also applies to running a stand-alone Python script that uses tools. If a license is unavailable, a tool will fail and return an error message. Higher license levels provide access to a greater number of tools. For example, if you have an ArcGIS for Desktop Basic license and attempt to run a tool that is part of ArcGIS for Desktop Advanced only, the tool will fail to run.

A tool from an ArcGIS extension, such as ArcGIS 3D Analyst for Desktop or ArcGIS Spatial Analyst for Desktop, requires an additional license for that extension. Thus, if you do not have an ArcGIS Spatial Analyst license and attempt to run a tool that is part of the Spatial Analyst toolbox, the tool will fail to run. For example, in the following code, the Slope function from the Spatial Analyst module is called with a raster digital elevation model (DEM) as input:

```
import arcpy
arcpy.sa.Slope("C:/Data/dem", "DEGREE")
```

If no Spatial Analyst license is available, the following error is generated:

```
ERROR 000824: The tool is not licensed.
Failed to execute (Slope).
```

Every tool checks to ensure that it has the proper license. To avoid having a script fail because of an unlicensed tool, it is a good practice to check for licenses at the beginning of the script.

Licenses for the following six products can be checked:

1. `arcview`

2. `arceditor`

3. `arcinfo`

4. `engine`

5. `enginegeodb`

6. `arcserver`

The product level can be set by importing the appropriate product module prior to importing ArcPy. For example, to set the desktop product license level to ArcGIS for Desktop Basic (formerly ArcView), a script would start with the following code:

```
import arcview
import arcpy
```

Notice that Python still references the original license levels.

The desktop product level cannot be set once ArcPy is imported. If a license is not explicitly set, the license is initialized based on the highest available license level the first time ArcPy is imported. In general, therefore, there is no need to set the product level in Python, and you will not see many scripts that include it.

*Note: The setting of the product and extensions is necessary only within stand-alone scripts. If you are running tools from the Python window or using script tools, the product is already set from within the application, and the active extensions are based on the Extensions dialog box.*

The CheckProduct function can be used to check whether the requested license is available. For example, the following code determines whether an ArcGIS for Desktop Advanced (formerly ArcInfo) license is available:

```
if arcpy.CheckProduct("arcinfo") == "Available":
```

*Note: The only parameter for* CheckProduct *is a single string, which should be one of the six product codes listed previously. It is not case sensitive, so "arcinfo" is the same in Python as "ArcInfo."*

The result of the CheckProduct function is a string, which can have one of five possible values:

1. AlreadyInitalized—license has already been set in the script.

2. Available—requested license is available to be set.

3. Unavailable—requested license is unavailable to be set.

4. NotLicensed—requested license is not valid.

5. Failed—system failure occurred during the request.

The ProductInfo function reports what the current product license is, as follows:

```
import arcpy
print arcpy.ProductInfo()
```

The ProductInfo function returns a string value that has the value NotInitialized if no license has been set yet, or else it returns the current product license.

Licenses for extensions can be retrieved for use in a script and returned once they are no longer needed. This is analogous to checking out licenses from within ArcMap or ArcCatalog using the Customize > Extensions option. The CheckExtension function is used to check whether a license is available to be checked out. For example:

```
import arcpy
arcpy.CheckExtension("spatial")
```

*Note: Similar to the product codes, license names are not case sensitive.*

The CheckExtension function returns a string and can have one of four possible values:

1. Available—requested license is available to be set.

2. Unavailable—requested license is unavailable to be set.

3. NotLicensed—requested license is not valid.

4. Failed—system failure occurred during license request.

Once the availability of a license is determined, the CheckOutExtension function can be used to actually obtain the license. Once a script runs the tools that required the particular license, the CheckInExtension function can be used to return the license to the license manager. For example, the following code first checks the availability of a license for ArcGIS 3D Analyst, and if the license is available, a license is obtained and then returned after the tool is finished running:

```python
import arcpy
from arcpy import env
env.workspace = "C:/Data"
if arcpy.CheckExtension("3D") == "Available":
    arcpy.CheckOutExtension("3D")
    arcpy.Slope_3d("dem", "slope", "DEGREES")
    arcpy.CheckInExtension("3D")
else:
    print "3D Analyst license is unavailable."
```

The CheckOutExtension function returns a string, and there are three possible return values: (1) NotInitialized, (2) Unavailable, and (3) CheckedOut. Typically, you would use the CheckExtension function to first determine the availability of the license before using the CheckOutExtension function. The CheckInExtension function returns a string, and there are three possible return values: (1) NotInitialized, (2) Failed, and (3) CheckedIn.

# 5.12 Accessing ArcGIS Desktop Help

The ArcGIS Help Library contains separate sections on using Python for geoprocessing and using the ArcPy site package. In ArcMap on the menu bar, click Help > ArcGIS Desktop Help. Or on the taskbar, click the Start button, and then, on the Start menu, click All Programs > ArcGIS > ArcGIS for Desktop Help > ArcGIS 10.1 for Desktop Help, and under the Contents tab, click Geoprocessing > Python. This documentation contains explanations of the basics of Python and how to carry out geoprocessing tasks in ArcGIS for Desktop using Python code.

Also under the Contents tab, click Geoprocessing > ArcPy for the ArcPy site package. All ArcPy functions and classes are listed and described in detail, and sample code is provided. There are also separate sections on the Data Access, Mapping, Network Analyst, and Spatial Analyst modules. ➜

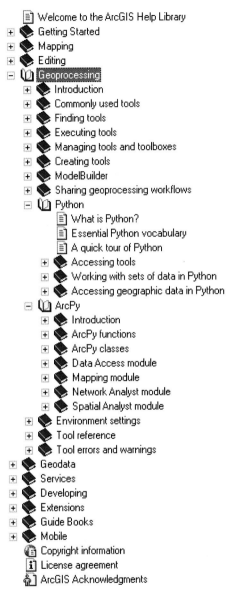

All the Help pages contain sample code. For example, the Help page for the `Exists` function (under ArcPy > ArcPy functions > General data functions) looks like the example in the figure.

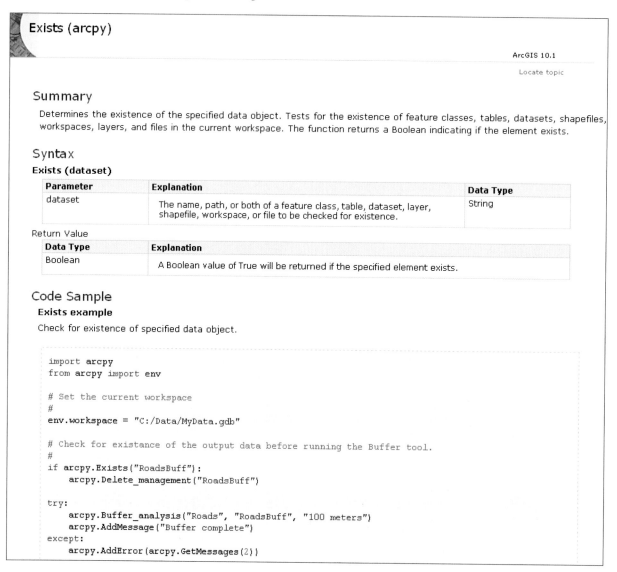

### Exists (arcpy)

ArcGIS 10.1

Locate topic

#### Summary

Determines the existence of the specified data object. Tests for the existence of feature classes, tables, datasets, shapefiles, workspaces, layers, and files in the current workspace. The function returns a Boolean indicating if the element exists.

#### Syntax

**Exists (dataset)**

| Parameter | Explanation | Data Type |
|---|---|---|
| dataset | The name, path, or both of a feature class, table, dataset, layer, shapefile, workspace, or file to be checked for existence. | String |

Return Value

| Data Type | Explanation |
|---|---|
| Boolean | A Boolean value of True will be returned if the specified element exists. |

#### Code Sample

**Exists example**

Check for existence of specified data object.

```python
import arcpy
from arcpy import env

# Set the current workspace
#
env.workspace = "C:/Data/MyData.gdb"

# Check for existance of the output data before running the Buffer tool.
#
if arcpy.Exists("RoadsBuff"):
    arcpy.Delete_management("RoadsBuff")

try:
    arcpy.Buffer_analysis("Roads", "RoadsBuff", "100 meters")
    arcpy.AddMessage("Buffer complete")
except:
    arcpy.AddError(arcpy.GetMessages(2))
```

The sample code assumes familiarity with Python scripting but typically provides good examples of how a particular function (or class) is used. You can copy all or parts of the code and paste it into a Python editor.

Individual geoprocessing tools also include Help with an explanation of how the tool works, as well as the tool syntax and sample Python code. Help can be accessed in several ways: (1) in ArcToolbox, right-click a tool and click Help; (2) on a tool dialog box, click the Show Help button to show the Help panel for the tool, and then click the Tool Help button; and (3) in ArcGIS 10.1 Help, under the Contents tab, click Geoprocessing > Tool reference—this allows you to navigate to a tool's Help page using the same structure of toolboxes, toolsets, and tools as in ArcToolbox.

For example, the Help page for the Copy tool in the Data Management toolbox (under the General toolset) looks like the example in the figure.

## Copy (Data Management)

ArcGIS 10.1

Locate topic

License Level:  ☑ Basic   ☑ Standard   ☑ Advanced

### Summary

Copies input data and pastes the output to the same or another location regardless of size. The data type of the Input and Output Data Element is identical.

### Usage

- If a feature class is copied to a feature dataset, the spatial reference of the feature class and the feature dataset must match; otherwise, the tool fails with an error message.
- Any data dependent on the input is also copied. For example, copying a feature class or table that is part of a relationship class also copies the relationship class. The same applies to a feature class that has feature-linked annotation, domains, subtypes, and indices—all are copied along with the feature class. Copying geometric networks, network datasets, and topologies also copies the participating feature classes.
- Copying a mosaic dataset copies the mosaic dataset to the designated location; the images referenced by the mosaic dataset are not copied.

### Syntax

Copy_management (in_data, out_data, {data_type})

| Parameter | Explanation | Data Type |
|---|---|---|
| in_data | The data to be copied to the same or another location. | Data Element |
| out_data | The name for the output data. | Data Element |
| data_type (Optional) | The type of the data to be renamed. The only time you need to provide a value is when a geodatabase contains a feature dataset and a feature class with the same name. In this case, you need to select the data type (feature dataset or feature class) of the item you want to rename. | String |

### Code Sample

#### Copy example 1 (Python window)

The following Python window script demonstrates how to use the Copy function in immediate mode.

```
import arcpy
from arcpy import env

env.workspace = "C:/data"
```

The syntax is provided, with a detailed explanation of each parameter. The code samples are typically quite short but demonstrate the specific use of the tool.

## Points to remember

- The ArcPy site package introduced in ArcGIS version 10 provides access to the Python geoprocessing functionality in ArcGIS. It is the successor to the `ArcGISscripting` module from earlier versions. ArcPy is organized in modules, functions, and classes.

- All geoprocessing tools in ArcGIS are provided as functions. Once ArcPy is imported to a Python script, you can run all the geoprocessing tools found in the standard toolboxes that are installed with ArcGIS. The syntax for running a tool is `arcpy.<toolname_toolboxalias>(<parameters>)`. The documentation on each tool provides details on the required and optional parameters needed for a tool to run. Additional nontool functions in ArcPy are available to support geoprocessing tasks.

- Classes in ArcPy are used to create objects. Commonly used classes are the `env` class and the `SpatialReference` class. The syntax for setting the property of a class is

      arcpy.<classname>.<property> = <value>.

- Messages that result from running a tool can be retrieved using message functions, including `GetMessages`, `GetMessage`, and `GetMaxSeverity`. Messages can consist of information, warning, or error messages.

- Several functions are available to check available licenses for products and extensions, to check out licenses, and to check licenses back in.

- ArcGIS for Desktop Help contains many examples of Python code, including the Help page for individual geoprocessing tools.

# Chapter 6
## Exploring spatial data

## 6.1 Introduction

This chapter describes several approaches to exploring spatial data, including checking for the existence of datasets and describing datasets in a workspace. List functions can be used not only to list datasets, but also to list elements such as workspaces, fields, and tables. Lists are common in scripts because they make it possible to iterate over a large number of elements. The built-in Python functions can be used to manipulate lists. Tuples and dictionaries are also introduced in this chapter as additional data structures.

## 6.2 Checking for the existence of data

To work with geoprocessing tools in ArcGIS for Desktop applications, you typically specify input datasets on the tool dialog box by selecting from a list of available layers or by browsing to the correct dataset. Tool dialog boxes have built-in mechanisms to validate tool parameters. For example, if you enter the name of a dataset that does not exist, an error message will appear next to the parameter on the tool dialog box. In Python scripts, however, you also need to determine whether datasets do, in fact, exist. This can be done using the `Exists` function. This function returns a Boolean value, indicating whether the element exists or not. The `Exists` function can be used to determine the existence of feature classes, tables, datasets, shapefiles, workspaces, layers, and other files.

The syntax of the `Exists` function is

```
arcpy.Exists(<dataset>)
```

For example, the following code determines whether a particular shapefile exists:

```
import arcpy
print = arcpy.Exists("C:/Data/streams.shp")
```

The returned value (`True` or `False`) is printed to the screen.

When working with Python, you should keep two types of paths in mind:

1. System paths—these are the paths recognized by the Windows operating system.

2. Catalog paths —these are the paths that only ArcGIS recognizes.

For example, C:\Data and C:\Data\streams.shp are system paths that are recognized by Windows and therefore also by Python. On the other hand, C:\Data\study.gdb\final\streets is a catalog path used in ArcGIS for Desktop applications. In this example, study.gdb is a file geodatabase, final is a feature dataset, and streets is a feature class. Such catalog paths are recognized by ArcGIS, but not by Windows or by the built-in Python functions. The existence of a path can be determined in Python using a built-in Python function such as `os.path.exists`. However, built-in Python functions do not work with catalog paths. To work with catalog paths in Python, you need to use ArcPy functions such as `Exists`.

A few other notes are in order pertaining to catalog paths:

- A file geodatabase is referred to as .gdb—for example, study.gdb.

- A personal geodatabase is referred to as .mdb—for example, study.mdb.

- An enterprise geodatabase is referred to as .sde—for example, study.sde.

Catalog paths consist of two parts: the workspace and the base name. In the preceding example, C:\Data\study.gdb\final\ is the workspace and streets is the base name. System paths also contain a base name, which is the last part of the path that does not contain a slash. For example, in C:\Data\streams.shp, the base name is streams.shp.

Working with elements inside a geodatabase presents some challenges. For example, how does ArcPy interpret a string like `C:\Data\study.gdb\final`? Is `final` interpreted as a feature class, a feature dataset, a raster, a table? The term is context specific. For example, if the string is used to set the current workspace, `final` is considered the name of a feature dataset, but if the string is used to specify an output feature class, `final` is considered the name of the feature class. In the next section, you will see how to determine the type of element being used.

# 6.3 **Describing data**

Geoprocessing tools work with all types of geospatial data. Each data
type has properties that can be used to control the flow of a script. The
`Describe` function can be used to determine the properties of the input
feature class, including the feature type (point, polyline, polygon, and
others). The properties of a dataset are often needed to validate tool parameters. For example, with the Clip tool, the Clip Features parameter has to be
polygons. Tool dialog boxes have built-in validation to prevent the selection
of feature classes that are an invalid feature type. In a Python script, you
can use the `Describe` function to determine the feature type of a dataset
before using it in a tool.

The syntax to describe a dataset is as follows:

```
import arcpy
<variable> = arcpy.Describe(<input dataset>)
```

Running the code returns an object that contains the properties of the dataset. These properties can be accessed using an `<object>.<property>`
statement. For example, the following code describes a shapefile, and then
prints the geometry shape type:

```
import arcpy
desc = arcpy.Describe("C:/Data/streams.shp")
print desc.shapeType
```

The `Describe` function can be used on various datasets, including geodatabase feature classes, shapefiles, rasters, and tables. The properties of
`Describe` are dynamic, meaning that the properties available depend on
the data type being described.

Because there are so many different properties, they are organized into
a series of property groups. This grouping does not really affect the syntax—the name of the property group does not appear in the syntax—but it
provides a logical organization of all the properties.

One property group is `FeatureClass`. It includes the `shapeType`
property used in the preceding example code. The `FeatureClass` properties provide access to the `FeatureClass` data type, which can be a
geodatabase, shapefile, coverage, or other type of feature class.

You have already seen how to access properties using an
`<object>.<property>` statement. In many cases, you need to access
properties to be able to specify tool parameters correctly. For example, the
following code determines the geometry shape type of the clip features, and
the Clip tool is run only if the shape type is polygons:

```
import arcpy
arcpy.env.workspace = "C:/Data"
infc = "streams.shp"
clipfc = "study.shp"
outfc = "streams_clip.shp"
desc = arcpy.Describe(clipfc)
type = desc.shapeType
if type == "Polygon":
    arcpy.Clip_analysis(infc, clipfc, outfc)
else:
    print "The clip features are not polygons."
```

There are many other useful property groups in the `Describe` function—for example, the `Dataset` properties. This includes properties
such as the type of dataset and the spatial reference. The following code
describes a shapefile and prints the type of dataset and the name of the
spatial reference:

```
import arcpy
fc = "C:/Data/streams.shp"
desc = arcpy.Describe(fc)
sr = desc.spatialReference
print "Dataset type: " + desc.datasetType
print "Spatial reference: " + sr.name
```

In the case of a shapefile, it is clear that the dataset type is a feature class.
However, in many cases, the dataset type is not so obvious. For example,
the following code determines the dataset type of an element inside a
geodatabase:

```
import arcpy
element = "C:/Data/study.gdb final"
desc = arcpy.Describe(element)
print "Dataset type: " + desc.datasetType
```

Determining the data type can be important because elements inside a geodatabase do not have file extensions. An element called final could be a feature
class, a feature dataset, a geodatabase table, or any other valid data type.
There are numerous valid dataset types, including FeatureClass, FeatureDa-
taset, RasterBand, RasterDataset, and Table. A complete list of valid dataset
types can be found in ArcGIS Desktop Help for the `Describe` function.

The `Describe` function also returns a number of properties for `Describe` objects. It includes such properties as the file path, catalog path, name, file name, and base name. Running the following code prints a number of these different properties:

```
import arcpy
arcpy.env.workspace = "C:/Data/study.gdb"
element = "roads"
desc = arcpy.Describe(element)
print "Data type: " + desc.dataType
print "File path: " + desc.path
print "Catalog path: " + desc.catalogPath
print "File name: " + desc.file
print "Base name: " + desc.baseName
print "Name: " + desc.name
```

The result is as follows:

```
Data type: FeatureClass
File path: C:/Data/study.gdb
Catalog path: C:/Data/study.gdb/roads
File name: roads
Base name: roads
Name: roads
```

Paths and names can be a bit confusing, so these properties can be useful to confirm their values for a particular object, especially for elements inside a geodatabase.

Additional examples of using the `Describe` function on rasters are covered in chapter 9.

## 6.4  Listing data

Batch processing is one of the primary reasons for developing geoprocessing scripts. In chapter 2, you saw several ways to use batch processing, including the option to run tools from ArcToolbox in batch mode. However, there are limitations, and scripting provides a more powerful and flexible framework for batch processing.

One of the key tasks in batch processing is to make an inventory of the available data so a script can iterate over the data during processing. ArcPy includes a number of functions to inventory data. These functions return a Python list, which consists of a list of values. A list can contain any type of data, but strings are the most common when working with ArcPy list functions.

The principle behind using a list is that a script can loop through the values in the list to work with each one individually. Working with a list in Python typically requires a `for` loop statement. The ArcPy list functions include `ListFields`, `ListIndexes`, `ListDatasets`, `ListFeatureClasses`, `ListFiles`, `ListRasters`, `ListTables`, `ListWorkspaces`, and `ListVersions`.

The parameters of these functions are similar. Some functions require an input dataset because the items of interest are located within a dataset or object. Other functions do not require a dataset because the items of interest are located within the current workspace. A workspace is required because you cannot pass a path to these functions. All functions have a wild card (*), which defines a name filter. This filter is used to list only the elements that meet a particular criterion.

The `ListFeatureClasses` function returns a list of feature classes in the current workspace. The syntax of the function is

```
ListFeatureClasses ({wild_card}, {feature_type}, {feature_dataset})
```

This function has three parameters, all of which are optional. These parameters make it possible to limit the list by name, feature type, or feature dataset. For example, the following code returns a list of all feature classes in the current workspace:

```
import arcpy
from arcpy import env
env.workspace = "C:/Data"
fclist = arcpy.ListFeatureClasses()
```

All the list functions return a Python list. To confirm the contents of a list, you can print the values using a statement like the following:

```
print fclist
```

A printed list of feature classes looks something like the following example:

```
[u'floodzone.shp', u'roads.shp', u'streams.shp', u'wetlands.shp', ➜
   ➜ u'zipcodes.shp']
```

A Python list of elements is surrounded by square brackets ([ ]), and the elements are separated by a comma (,). In the case of a list of feature classes, the elements are strings.

The wild card can be used to limit the list by name. For example, the following code creates a list of all feature classes in the current workspace that start with the letter w:

```
fclist = arcpy.ListFeatureClasses("w*")
```

*Note: Remember that string values preceded by the letter u are Unicode strings. Unicode strings work just like regular strings but are more robust when working with different international character sets.*

The second parameter in the `ListFeatureClasses` function is the feature type. This parameter can be used to restrict the list to match certain data properties, such as point feature classes only. For example, the following code creates a list of all the point feature classes in the current workspace:

```
fclist = arcpy.ListFeatureClasses("", "point")
```

Notice the use of an empty string (`""`) for the wild card. Parameters have to be entered in the order defined by the syntax and cannot be skipped unless you specifically refer to them by name. Using `"*"` would give the same result, in this case. Valid feature types include annotation, arc, dimension, edge, junction, label, line, multipatch, node, point, polygon, polyline, region, route, and tic.

To create a list of raster datasets in the current workspace, you can use the `ListRasters` function. The syntax is very similar to the `ListFeatureClasses` function:

```
ListRasters ({wild_card}, {raster_type})
```

The two parameters are optional and allow you to restrict the list by name or type. The following code creates a list of all raster datasets in the current workspace:

```
import arcpy
from arcpy import env
env.workspace = "C:/Data"
rasterlist = arcpy.ListRasters()
```

To restrict the list to all rasters that are TIFF images, you can specify the raster type parameter as follows:

```
rasterlist = arcpy.ListRasters("", "tif")
```

A complete list of supported raster data types can be found in ArcGIS Desktop Help. The raster type parameter should be specified as the file extension, not the generic name of the format. For example, a TIFF file has the file extension .tif, so `tif` is the correct syntax. Similarly, a JPEG file has the file extension .jpg, so `jpg` is the correct syntax. The syntax for raster data types is not case sensitive, so both `TIF` and `tif` are correct. Also note that the Esri GRID format does not have a file extension and the proper syntax for this format is `GRID` (also not case sensitive).

Another list function is `ListFields`. This function lists the fields in a feature class or table in a specified dataset. The syntax is

```
ListFields(dataset, {wild_card}, {field_type})
```

The `ListFields` function has three parameters—name, field type, and dataset—of which the dataset is required. The dataset is the specified features class or table whose fields will be returned as values in a list. For example, the following code creates a list of all the fields in a shapefile:

```
import arcpy
from arcpy import env
env.workspace = "C:/Data"
fieldlist = arcpy.ListFields("roads.shp")
```

The two optional parameters allow you to restrict the list by name or field type. The following code creates a list of all the fields in a shapefile that are integers:

```
fieldlist = arcpy.ListFields("roads.shp", "", "Integer")
```

Valid field types include All, BLOB (binary large object), Date, Double, Geometry, GUID (globally unique identifier), Integer, OID (object identifier), Raster, Single, SmallInteger, and String. Field types are strings and are not case sensitive.

The `ListFields` function returns a list of field objects. By contrast, most of the other list functions return a list of strings. Field object properties include the field name, alias, type, and length. For example, the following script creates a list of fields of type String and determines for each text field what the length of the field is:

```
import arcpy
from arcpy import env
env.workspace = "C:/Data"
fieldlist = arcpy.ListFields("roads.shp", "", "String")
for field in fieldlist:
    print field.name + " " + str(field.length)
```

The `length` property returns an integer and is converted to a string for printing purposes.

# 6.5  **Using lists in for loops**

Once you have the list of desired values, you can use the list for batch processing. This is most commonly accomplished using a for loop. A for loop
can be used to iterate over the list, one element at a time, and when there
are no values left to be iterated, the loop is finished. For example, consider
using a for loop with raster data. The code is as follows:

```
import arcpy
from arcpy import env
env.workspace = "C:/Data"
tifflist = arcpy.ListRasters("", "TIF")
for tiff in tifflist:
    arcpy.BuildPyramids_management(tiff)
```

In this example, the ListRasters function is used to create a list of TIFF
files. The for loop iterates over each element in the list and builds pyramids for each. This becomes quite powerful because it can automate rather
tedious tasks. For example, building pyramids for hundreds of rasters could
become quite time consuming. The few lines of Python code used here
carry out this task and the code is the same whether there are only a few
raster datasets or several hundred.

Iteration using a for loop can also be used in combination with the
ListFields function to provide a detailed description of the fields and
their properties. The following code creates a list of fields in a single shapefile, and then prints the name, type, and length of each field:

```
import arcpy
from arcpy import env
env.workspace = "C:/Data"
fieldlist = arcpy.ListFields("roads.shp")
for field in fieldlist:
    print "{0} is a type of {1} with a length of {2}".format(field.name, ➜
➜ field.type, field.length)
```

# 6.6 **Working with lists**

Lists are a versatile Python type and can be manipulated in many different
ways. So although a list of feature classes, fields, or rasters is created using
an ArcPy function, the list can be manipulated in Python using the functions and methods of a Python list. Following are a few examples.

The number of feature classes in a workspace can be determined using
the built-in Python `len` function. The code is as follows:

```
import arcpy
from arcpy import env
env.workspace = "C:/Data/study.gdb"
fcs = arcpy.ListFeatureClasses()
print len(fcs)
```

Lists can be sorted using the `sort` method. The default sorting is in alphanumerical order, but it can be reversed using the `reverse` argument of
the `sort` method. The following code creates a list of feature classes, sorts
them alphanumerically, and prints their names. The sorting is then reversed
and the names are printed again. The code is as follows:

```
import arcpy
from arcpy import env
env.workspace = "C:/Data/study.gdb"
fcs = arcpy.ListFeatureClasses()
fcs.sort()
print fcs
fcs.sort(reverse = True)
print fcs
```

The result is as follows:

```
[u'hospitals', u'parks', u'roads', u'streams', u'wetlands']
[u'wetlands', u'streams', u'roads', u'parks', u'hospitals']
```

Working with lists is covered in more detail in chapter 4, including the use
of indexing, slicing, and list methods. All of it can be used when working
with lists of feature classes, tables, fields, rasters, or other spatial data types.

# 6.7 Working with tuples

Lists are common in Python, and you will often use lists when writing geo-processing scripts, including lists of map documents, layers, feature classes, fields, and more. Lists are quite versatile since you can modify them in many ways, as discussed in the previous section. Sometimes, however, you may want to use a list without allowing its elements to be modified. That is where *tuples* come in. Tuples are sequences of elements, just like lists, but tuples are immutable, meaning they cannot be changed. The syntax of a tuple is simple—separate a set of values with commas (,), and you have a tuple. For example, the following code returns a tuple with five elements:

```
>>> 1, 2, 3, 4, 5
```

The result is (1, 2, 3, 4, 5).

How to get a tuple with only one element? Add a comma (,), even though there is no element following the comma:

```
>>> 6,
```

The result is (6).

Working with tuples is similar to working with lists. Elements in the tuple have an index value, which can be used to obtain specific elements of the tuple. For example:

```
>>> x = ("a", "b", "c")
>>> x[0]
```

The result is 'a'.

However, the sequence of elements cannot be modified. So list operations such as deleting, appending, removing, and others are not supported by tuples. The only methods that work on tuples are count and index because these methods do not modify the sequence of elements. Other operations can be applied but return a different tuple. Running the following code slices a tuple and returns a different tuple:

```
>>> x = ("a", "b", "c", "d", "e", "f", "g")
>>> x[2:5]
```

The result is ('c', 'd', 'e').

Notice that the slicing operation returns another tuple, not a list. You cannot modify the elements of a tuple, unlike the elements of a list.

If you cannot modify a tuple, why are tuples important? First, some built-in Python functions and modules return tuples—in which case, you have to deal with them. Second, tuples are often used in dictionaries, which are covered next.

**>>> TIP**

In general, if you are using a set of values that won't be modified, use a tuple instead of a list.

# 6.8  **Working with dictionaries**

Lists and tuples are useful for grouping elements into a structure, and the elements can be referred to by their index number, starting with zero (0). Working with index numbers works fine, but it has its limitations. Consider the example of the following list of cities:

```
cities = ["Austin", "Baltimore", "Cleveland", "Denver"]
```

Suppose you want to have a database that contains the state for each city. You can do this by creating a list as follows:

```
states = ["Texas", "Maryland", "Ohio", "Colorado"]
```

Because the index numbers correspond, you can access elements from one list by using the index number from the other list. For example, to get the state for Cleveland, you can use the following code:

```
>>> states[cities.index("Cleveland")]
```

The result is 'Ohio'.

This process is useful but cumbersome. Imagine lists that have a large number of elements. In addition, some states will have more than one city. Making only a minor edit to one of the lists could disrupt the entire sequence. You can use tuples to ensure no changes are made to the sequence, but that also has its limitations. What you really need is a lookup table that would work as follows:

```
>>> statelookup["Cleveland"]
```

The result is 'Ohio'.

A lookup table is commonly used to display information from one table based on a corresponding value in another table. A Table Join operation in ArcGIS is an example of using a lookup table. In Python, one way to implement a lookup table is to use a *dictionary*. Dictionaries consist of pairs of keys and their corresponding values. Pairs are also referred to as the *items* of the dictionary. A dictionary item consists of a key, followed by a colon (:), and then the corresponding value. Items are separated by a comma (,). The dictionary itself is surrounded by curly brackets ({ }).

The dictionary for the preceding example would look as follows:

```
>>> statelookup = {"Austin": "Texas", "Baltimore": "Maryland",
"Cleveland": "Ohio", "Denver": "Colorado"}
```

You can now use this dictionary to look up the state for each city:

```
>>> statelookup["Cleveland"]
```

The result is 'Ohio'.

The order in which the items were created in the dictionary does not matter. The dictionary can be modified, and as long as the pairs of keys and their corresponding values are intact, the dictionary will continue to function. Keep in mind when creating the dictionary that the keys have to be unique, but the values do not.

Dictionaries can be created and populated at the same time, as in the preceding statelookup example. You can also create an empty dictionary first using only curly brackets ({}), and then add items to it. Here is the code to create a new empty dictionary:

```
>>> zoning = {}
```

Items can be added using square brackets ([ ]) and an assignment statement as follows:

```
>>> zoning["RES"] = "Residential"
```

You can continue to add items to the dictionary. Elements are sorted in alphanumerical order based on the key. The code is as follows:

```
>>> zoning["IND"] = "Industry"
>>> zoning["WAT"] = "Water"
>>> print zoning
```

The result is as follows:

```
{'IND': 'Industry', 'RES': 'Residential', 'WAT': 'Water'}
```

Items can be modified using the same syntax. Setting the value using a key that is already in use overwrites the existing value. The code is as follows:

```
>>> zoning["IND"] = "Industrial"
>>> print zoning
```

The result is as follows:

```
{'IND': 'Industrial', 'RES': 'Residential', 'WAT': 'Water'}
```

Items can be deleted using square brackets ([ ]) and the keyword `del` as follows:

```
>>> del zoning["WAT"]
>>> print zoning
```

The result is `{'IND': 'Industry', 'RES': 'Residential'}`.

There are several dictionary methods. The `keys` method returns a list of all the keys in the dictionary, as follows:

```
>>> zoning.keys()
```

The result is `['IND', 'RES']`.

The `values` method returns a list of all the values in the dictionary, as follows:

```
>>> zoning.values()
```

The result is `['Industrial', 'Residential']`.

The `items` method returns a list of all items in the dictionary—that is, all key-value pairs, as follows:

```
>>> zoning.items()
```

The result is `[('IND', 'Industrial'), ('RES', 'Residential')]`.

Dictionaries are not common in ArcPy, but there is one function that uses them: `GetInstallInfo`. This function returns a Python dictionary that contains the information on the installation type properties. The general syntax of the `GetInstallInfo` function is

```
GetInstallInfo (install_name)
```

For example, the following code returns the installation information for the product being used in Python:

```
import arcpy
install = arcpy.GetInstallInfo()
for key in install:
    print "{0}: {1}".format(key, install[key])
```

The `GetInstallInfo` function returns a dictionary, and the `for` loop prints all the pairs of keys and their corresponding values. The result looks something like the following:

```
SourceDir: D:\Desktop\
InstallDate: 6/1/2012
InstallDir: C:\Program Files\ArcGIS\Desktop10.1\
ProductName: desktop
BuildNumber: 2414
InstallType: N/A
Version: 10.0
SPNumber: 2
Installer: Paul
SPBuild: 10.0.2.3200
InstallTime: 13:34:26
```

## Points to remember

- The `Exists` function can be used to determine whether a particular dataset exists. The `Describe` function can be used to describe the properties of a dataset. These functions are commonly used to ensure that inputs for a script conform to expectations.

- List functions are used to facilitate batch processing. Once a list of elements is created, a script can be designed to iterate over all the elements in the list. For example, the `ListFeatureClasses` function can be used to create a list of all feature classes in a workspace, and a `for` loop can be used to iterate over all the elements in the list to perform the same operation on each feature class. List functions exist for different types of elements, including workspaces, files, datasets, feature classes, fields, rasters, tables, and others. Lists are very common in scripts.

- Tuples and dictionaries are important data structures in Python. A tuple is a sequence of elements, just like a list, but the elements of a tuple cannot be changed. A dictionary consists of pairs of keys and their corresponding values. Dictionaries work the same way as lookup tables.

# Chapter 7
## Manipulating spatial data

## 7.1  Introduction

This chapter introduces the ArcPy data access module `arcpy.da` for
working with data. This module allows control of an edit session, editing operations, cursor support, functions for converting tables and feature
classes to and from NumPy arrays for additional processing, and support
for versioning and replica workflows. This chapter focuses on cursors,
which are used to iterate over rows in a table. Different types of cursors
can be used to search records, add new records, and make changes to existing records. Search cursors can be used to carry out SQL query expressions
in Python. Validation of text and field names is also covered.

## 7.2  Using cursors to access data

In chapter 6, you saw how list functions can be used to iterate over a set
of values in a list, including feature classes, tables, and fields. A similar
approach using a *cursor* is used to work with the rows in a table. A cursor is
a database technology term for accessing a set of records in a table. Concep-
tually, it is similar to the way list functions work, where a cursor is used to
manipulate a list of records. A cursor can be used to iterate over the set of
records in a table or to insert new records into a table. Records in a table
are also referred to as rows. In ArcGIS, cursors can be used to read and
write geometries to and from records, row by row.

There are three types of cursors: search, insert, and update. These cursors have the following characteristics:

- A search cursor is used to retrieve rows.

- An insert cursor is used to insert rows.

- An update cursor is used to update and delete rows.

Each type of cursor is created by the corresponding function of the `arcpy.da` module: `SearchCursor`, `InsertCursor`, and `UpdateCursor`. All three cursors can work on a table, a table view, a feature class, or a feature layer. Table 7.1 describes the cursor methods. All three cursor functions create a cursor object that can be used to access row objects. The methods supported by the cursor object depend on the type of cursor created.

**Table 7.1  Methods supported by cursor type**

| Cursor Type | Method | Description |
|---|---|---|
| Search | next | Retrieves the next row |
| | reset | Resets the cursor to its starting position |
| Insert | insertRow | Inserts a row into the table |
| | next | Retrieves the next row object |
| Update | deleteRow | Removes the row from the table |
| | next | Retrieves the next row object |
| | reset | Resets the cursor to its starting position |
| | updateRow | Updates the current row |

*Note: The data access module was added with ArcGIS 10.1. Previously existing cursors, such as* `arcpy.InsertCursor`, *are still functional and valid, but the new* `arcpy.da` *cursors feature much faster performance.*

Cursors navigate in a forward direction. If a script needs to make multiple passes over the data by iteration, the cursor function must be run again. Search and update cursors can iterate using a `for` loop or in a `while` loop using the cursor's `next` method to move on to the next row. For a table with *n* rows, a script needs to make *n* calls to this `next` method. When the last row is reached, a call to the `next` method will return a `StopIteration` exception.

All three cursors have two required arguments: an input table and a list (or tuple) of field names. Search and update cursors also have several optional arguments. The general syntax for each cursor is as follows:

```
arcpy.da.InsertCursor(in_table, field_names)

arcpy.da.SearchCursor(in_table, field_names, {where_clause}, {spatial_ ➔
➔ reference}, {explore_to_points})

arcpy.da.UpdateCursor(in_table, field_names, {where_clause}, {spatial_ ➔
➔ reference}, {explore_to_points})
```

Each row retrieved from a table by using a cursor is returned as a list of field values, in the same order as specified by the field_name argument.

Following is an example of using a search cursor to iterate over the rows in a table. In the example code that follows, the SearchCursor function retrieves the rows in a table. A for loop is used to iterate over the rows in the table and print the values for a given field.

```
import arcpy
fc = "C:/Data/study.gdb/roads"
cursor = arcpy.da.SearchCursor(fc, ["STREETNAME"])
for row in cursor:
    print "Streetname = {0}".format(row[0])
```

Consider the attribute table of the roads feature class, as shown in the figure. ➔

The preceding script would result in a printout that is as follows:

```
Streetname = MARKHAM WOOD
Streetname = MARKHAM WOOD
Streetname = LAKE MARY
Streetname = AVENUE H
Streetname = FLORIDA
Streetname = LONGWOOD HILLS
Streetname = CENTRAL
Streetname = MYRTLE
Streetname = 434
. . .
```

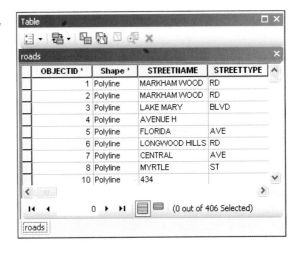

Working with values is revisited later in this chapter.

Search and update cursors also support `with` statements. A `with` statement has the advantage of guaranteeing closure and release of database locks and resetting iteration, regardless of whether the cursor finished running successfully or not. The previous example using a `with` statement is coded as follows:

```
import arcpy
fc = "C:/Data/study.gdb/roads"
with arcpy.da.SearchCursor(fc, ["STREETNAME"]) as cursor:
    for row in cursor:
        print "Streetname = {0}".format(row[0])
```

The following code illustrates how a new row is inserted. A cursor object is created using the `InsertCursor` function. Once the cursor is created, the `insertRow` method is used to insert a list of values that will be included by the new row.

The code is as follows:

*Note: The `InsertCursor` function is capitalized whereas the `insertRow` method is not. This style is adopted throughout ArcPy.*

```
import arcpy
fc = "C:/Data/study.gdb/roads"
cursor = arcpy.da.InsertCursor(fc, ["STREETNAME"])
cursor.insertRow(["NEW STREET"])
```

You can insert multiple rows using a loop:

```
cursor = arcpy.da.InsertCursor(fc, ["STREETNAME"])
x = 1
while x <= 5:
    cursor.insertRow(["NEW STREET"])
    x += 1
```

By default, the new rows are inserted at the bottom of the table. Fields in the table that are not included in the cursor are assigned the field's default value, typically "null" (this will vary with the database format).

The update cursor is used to update or delete the row at the current position of the cursor. The `updateRow` method is used to update the row. After a row object is retrieved from the cursor, the row is modified as needed and the modified row is passed to the table by calling the `updateRow` method.

In the following example, the cursor object is created using the `UpdateCursor` function. In the `for` loop, the values of one field (Acres) in the row are updated using the values of another field (Shape_Area). If the Shape_Area field is in square feet, the value needs to be divided by 43,560 to obtain the area in acres. The row is updated using the `updateRow` method on the row object.

```
import arcpy
fc = "C:/Data/study.gdb/zones"
cursor = arcpy.da.UpdateCursor(fc, ["ACRES", "SHAPE_AREA"])
for row in cursor:
    row[0] = row[1] / 43560
    cursor.updateRow([row])
```

The `deleteRow` method is used to delete the row at the current position of the `UpdateCursor`. After the cursor retrieves the row object, calling the `deleteRow` method deletes the row, as follows:

```
import arcpy
fc = "C:/Data/study.gdb/roads"
cursor = arcpy.da.UpdateCursor(fc, ["STREETNAME"])
for row in cursor:
    if row[0] == "MAIN ST":
        cursor.deleteRow()
```

Insert and update cursors support editing operations. In the ArcGIS geo-processing framework, a lock is set on the table when the cursor object is created. Locks prevent multiple processes from changing the same table at the same time. There are two types of locks: shared and exclusive. A shared lock is applied when a table or dataset is accessed. For example, opening a feature class in ArcMap and performing a query will result in a shared lock on the dataset. Multiple shared locks can exist, but no exclusive locks are permitted if there is already a shared lock. An exclusive lock is applied when changes are being made to a table or dataset. Examples include editing and saving a feature class in ArcMap, changing the schema of a table or feature class in ArcCatalog, and using an insert or update cursor on a feature class in Python.

Insert and update cursors apply an exclusive lock to a table or feature class, preventing other processes from making changes. In addition, insert and update cursors cannot be applied if an exclusive lock already exists. For example, two scripts cannot simultaneously create an insert or update cursor on the same dataset.

Once an exclusive lock is created by a script, the lock persists until the script releases the lock. This is accomplished using the `del` statement to delete the cursor object creating the lock. Without this statement, a lock could needlessly block other applications or scripts from accessing a dataset. A typical script that creates an insert or update cursor should therefore include two `del` statements—one to delete the row object, such as `del row`, and one to delete the cursor object, such as `del cursor`. For example:

```
import arcpy
fc = "C:/Data/study.gdb/roads"
cursor = arcpy.da.UpdateCursor(fc, ["STREETNAME"])
for row in cursor:
    if row[0] == "MAIN ST":
        cursor.deleteRow()
del row
del cursor
```

**>>> TIP**

Forgetting to use the `del` statements at the end of a script can lead to errors so be sure to include the `del` statements after using insert and update cursors. Search and update cursors support `with` statements, which guarantee closure and release of database locks. Therefore, when a `with` statement is used, the `del` statements are not needed.

## 7.3 Using SQL in Python

One of the most common processing steps in geoprocessing is to apply a query using Structured Query Language (SQL). SQL is used to define one or more criteria based on attributes, operators, and calculations. For example, SQL is used in the Select By Attributes dialog box in ArcMap, as well as in many different tools in ArcToolbox, including the Select tool.

***Note:*** *It is assumed here that you are familiar with writing SQL expressions. The basics of working with SQL can be found in ArcGIS Desktop Help, under the topic "Building a query expression."*

SQL queries can be carried out in Python using the `SearchCursor` function. The syntax for `SearchCursor` is

```
SearchCursor(in_table, field_names {where_clause}, {spatial_reference}, ➜
➜ {fields}, {explode_to_points})
```

The optional `where_clause` parameter consists of an SQL expression. SQL syntax varies slightly depending on the dataset being queried, and the same syntax rules apply when SQL is used in the search cursor parameters.

The following code shows an example of using an SQL expression:

```
import arcpy
fc = "C:/Data/study.gdb/roads"
cursor = arcpy.da.SearchCursor(fc, ["NAME", "CLASSCODE"], '"CLASSCODE" = ➡
➡ 1')
for row in cursor:
    print row[0]
del row
del cursor
```

SQL syntax can be cumbersome since it depends on the format of the feature class. For example, field delimiters for shapefiles and file geodatabase feature classes consist of double quotation marks (" "), personal geodatabase feature classes use square brackets ([ ]), and ArcSDE geodatabase feature classes use no delimiters. To avoid confusion and to ensure the field delimiters are the correct ones, you can use the `AddFieldDelimiters` function. The syntax of this function is

```
AddFieldDelimiters(datasource, field)
```

The function recognizes the type of dataset being used, such as a shapefile or personal geodatabase, and adds the correct delimiters. In the following example code, the field name is first assigned to a variable and the `AddFieldDelimiters` function is used to add the correct delimiters to the field name prior to using it in an SQL expression:

```
import arcpy
fc = "C:/Data/zipcodes.shp"
fieldname = "CITY"
delimfield = arcpy.AddFieldDelimiters(fc, fieldname)
cursor = arcpy.da.SearchCursor(fc, ["NAME", "CLASSCODE"], delimfield + " ➡
➡ = 'LONGWOOD'")
for row in cursor:
    print row[0]
del row
del cursor
```

SQL is commonly used in other functions as well. A number of built-in tools use SQL—for example, the Select tool. The generic syntax of the Select tool is

```
Select_analysis(in_features, out_feature_class, {where_clause})
```

The `where_clause` parameter is an SQL expression, and to avoid confusion over the proper field delimiters, the following example code also uses the `AddFieldDelimiters` function:

```
import arcpy
infc = "C:/Data/zipcodes.shp"
fieldname = "CITY"
outfc = "C:/Data/zip_select.shp"
delimfield = arcpy.AddFieldDelimiters(infc, fieldname)
arcpy.Select_analysis(infc, outfc, delimfield + " = 'LONGWOOD'")
```

## 7.4  Working with table and field names

When working with many different datasets, it is important to be able to determine whether tables and fields have a valid and unique name. This is especially important when creating data in a geodatabase to avoid accidentally overwriting data. The `Exists` function covered in chapter 6 can be used to determine whether a specific table is unique to a given workspace.

The `ValidateTableName` function can be used to determine whether a specific table name is valid for a specific workspace. The syntax of this function is

```
ValidateTableName(name, {workspace})
```

The function takes a table name and a workspace and returns a valid table name for the workspace. If the table name is already valid, the function returns a string that is the same as the input. Names may have invalid characters, which are replaced by an underscore (_). Specific databases may also have words that are reserved and cannot be used in a table name.

For example, the following script determines whether `all roads` is a valid table name for the file geodatabase called study:

```
import arcpy
tablename = arcpy.ValidateTableName("all roads", "C:/Data/study.gdb")
print tablename
```

In this example, the value `all_roads` is returned for the table name. An underscore (_) is added because the table name cannot contain spaces. Validating table names is commonly used when moving data from one workspace to another. For example, the following code uses the Copy Features tool to copy all shapefiles from a folder to a geodatabase. The .shp file extension is removed from the file name using the `basename` property and the names are validated prior to copying the shapefiles to feature classes, as follows:

```
import arcpy
import os
from arcpy import env
env.workspace = "C:/Data"
outworkspace = "C:/Data/test/study.gdb"
fclist = arcpy.ListFeatureClasses()
for shapefile in fclist:
    fcname = arcpy.Describe(shapefile).basename
    newfcname = arcpy.ValidateTableName(fcname)
    outfc = os.path.join(outworkspace, newfcname)
    arcpy.CopyFeatures_management(shapefile, outfc)
```

A similar approach can be applied to field names using the `ValidateFieldName` function. Fields cannot be added unless their name is valid, so unless field names are validated first, a script may fail during execution. The syntax of this function is

```
ValidateFieldName(name, {workspace})
```

The `ValidateFieldName` function takes a field name and a workspace and returns a valid field name for the workspace. Invalid characters are replaced by an underscore (_). The following code validates the name of a new field, and then uses the returned value to add the new field using the Add Field tool:

```
import arcpy
fc = "C:/Data/roads.shp"
fieldname = arcpy.ValidateFieldName("NEW%^")
arcpy.AddField_management(fc, fieldname, "TEXT", "", "", 12)
```

In this example, the string `"NEW%^"` is replaced by `"NEW__"`.

*Note: Each invalid character is replaced by an underscore, so in this example string, there are two underscores.*

Similar to the `ValidateTableName` function, the `ValidateFieldName` function does not determine whether the field name exists so a script could still fail or an existing field could still be overwritten. To determine whether a field name exists, the `ListFields` function can be used to create a list of the fields in a table or feature class, and the new field name can be compared against this list.

An alternative to trying to determine whether a table name exists is to use the `CreateUniqueName` function. This function creates a unique name in the specified workspace by appending a number to the input name. This number is increased until the name is unique. For example, if the name "Clip" already exists, the `CreateUniqueName` function will change it to "Clip0"; if this name also exists, the function will change the name to "Clip1," and so on. This function can be used only to create unique table names within a workspace. It does not work for field names. For example:

```
import arcpy
from arcpy import env
env.workspace = "C:/Data"
unique_name = arcpy.CreateUniqueName("buffer.shp")
arcpy.Buffer_analysis("roads.shp", unique_name, "100 FEET")
```

When this code is run the first time and the file `buffer.shp` does not exist, the resulting feature class is called `buffer.shp`. When the code is then run a second time, the resulting feature class is called `buffer0.shp`.

You may have noticed that most geoprocessing tool dialog boxes use a similar approach to providing default names for output datasets. For example, when the Clip tool is run on a feature class, the output, by default, is the name of the input feature class followed by "_Clip," such as rivers_Clip. When the Clip tool is run again on the same feature class and the workspace has not been altered, the default name becomes rivers_Clip1.

## 7.5  Parsing table and field names

The ArcGIS geoprocessing environment always uses fully qualified names for feature classes and tables. For example, to process a feature class called roads, a geoprocessing tool needs to know not just the name, but also the path, the name of the database, and the owner of the data. This is essential when working with geodatabases.

The `ParseTableName` function can be used to split the fully qualified name for a dataset into its components. The syntax of this function is

```
ParseTableName(name, {workspace})
```

The `ParseTableName` function returns a single string with the database name, owner name, and table name, each separated by a comma (,). For example, the following code uses the `ParseTableName` function on a feature class to obtain the components of the fully qualified name, and then splits these components into a list:

```
import arcpy
from arcpy import env
env.workspace = "C:/Data/study.gdb"
fc = "roads"
fullname = arcpy.ParseTableName(fc)
namelist = fullname.split(", ")
databasename = namelist[0]
ownername = namelist[1]
fcname = namelist[2]
print databasename
print ownername
print fcname
```

*Note: Parsing is also possible using the parsing functions in Python. However, parsing functions in Python assume a specific syntax, which may change for a different dataset. For example, the parsing syntax would need to be different for each type of geodatabase (file, personal, SDE). The ArcPy parsing functions are specifically designed to work with geodatabases and are therefore more robust.*

A similar approach can be used to split the fully qualified name of a field in a table into its components using the `ParseFieldName` function. This function returns a single string with the database name, owner name, table name, and field name, each separated by a comma (,).

## 7.6 Working with text files

So far, most of the information data you have worked with in this book, such as paths, values, lists of values, and more, is located inside the script itself or inside data files in GIS format, such as shapefiles, geodatabases, and database tables. In many cases, information data is also located in plain text files. Python has several functions to work with text files in other formats. In many cases, these files come from other applications and Python can be used to manipulate these files for use in ArcGIS.

You can open files using the `open` function, which has the following syntax:

```
open(name, {mode}, {buffering})
```

The only required argument is a file name and the function returns an object. For example, the following code opens an existing text file (.txt) on disk:

```
>>> f = open("C:/Data/sample.txt")
```

Using only a file name as a parameter returns a file object you can read from. If you want to do something else, such as write to the file, this must be stated explicitly by specifying a mode. The most common values for the mode are as follows:

r:  read mode

w:  write mode

+:  read/write mode (added to another mode)

b:  binary mode (added to another mode—for example, rb, wb, and others)

a:  append mode

If no mode argument is provided, the read mode is used by default. Write mode allows you to write to the file. Read/write mode can be added to any of the other modes to indicate both reading and writing is allowed. Binary mode allows you to change the way a file is handled. By default, Python assumes you are dealing with text files, which contain characters. If you are working with some other kind of file such as an image, you can add *b* to the mode—for example, "rb". Append mode means that any data written to the file is automatically added to the end of the file.

The buffering parameter controls the buffering of the files. The default is unbuffered, which means all reads and writes go directly to and from disk. When buffering is set to True, Python may use memory instead of disk space to improve performance. For modest file sizes, buffering is normally not needed.

To create a new file object, you can use the open function and specify write mode:

```
>>> f = open("C:/Data/mytext.txt", "w")
```

Several file methods exist to manipulate the contents of a text file, including write, read, and close. Consider the following example:

```
>>> f = open("C:/Data/mytext.txt", "w")
>>> f.write("Geographic Information Systems")
>>> f.close()
```

Running this code creates a new file object. If the file mytext.txt already exists, the existing file will be overwritten, so be careful. The write method is used to write a string to the file and the close method is used to close the file (and save its contents).

Reading a file works as follows:

```
>>> f = open("C:/Data/mytext.txt")
>>> f.read()
```

The result is

```
'Geographic Information Systems'
```

When opening a file just to read its contents, it is not necessary to specify a mode because `r` (read mode) is the default. The `read` method is used to read the contents of the text file. When no argument is specified, the script reads the entire contents of the file. An optional argument can be supplied to indicate the number of characters to be read. For example:

```
>>> f = open("C:/Data/mytext.txt")
>>> f.read(3)
```

The result is

```
'Geo'
```

Once you open a file, reading the file is like making a single pass through the string of characters. The next time you use the `read` method, the script picks up where it stopped the last time. For example:

```
>>> f.read(5)
```

The result is

```
'graph'
```

For example:

```
>>> f.read()
```

The result is

```
'ic Information Systems'
```

After you start reading a file, you can use the `seek` method to set the file's current position without opening the file again. For example:

```
>>> f.seek(0)
>>> f.read(10)
```

The result is

```
'Geographic'
```

Most text files consist of multiple lines and there are several file methods to work with lines. You can read a single line with the `readline` method. To read all the lines of a file and have them returned as a list, you can use the `readlines` method.

Next, take a look at some examples. Consider the text file, as shown in the figure.

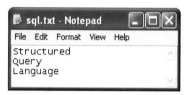

Reading the contents of this file can be accomplished in a number of ways, starting with the `read` method, as follows:

```
>>> f = open("C:/Data/sqltext.txt")
>>> f.read()
```

The result is

```
'Structured\nQuery\nLanguage'
```

Notice that the `read` method reads all the lines and returns the result as a single string. The line separators in the returned text file are the characters \n.

Next is the `readline` method. For example:

```
>>> f = open("C:/Data/sqltext.txt")
>>> f.readline()
```

The result is

```
'Structured\n'
```

And run again:

```
>>> f.readline()
```

The result is

```
'Query\n'
```

And run again:

```
>>> f.readline()
```

The result is

```
'Language'
```

The `readline` method reads the next line from the text file and returns it as a string. Continuing to use the `readline` method returns the lines that follow. The line separators (\n) are also returned.

Finally, the `readlines` method reads all the lines in the files and returns the result as a list. For example:

```
>>> f = open("C:/Data/sqltext.txt")
>>> f.readlines()
```

The result is

```
['Structured\n', 'Query\n', 'Language']
```

Writing a file with multiple lines can be accomplished using the `write` and `writelines` methods. To add new lines, the line separators (\n) have to be used. For example:

```
>>> f = open("C:/Data/tintext.txt", "w")
>>> f.write("Triangulated\nIrregular\nNode")
>>> f.close()
```

Running this code creates a new text file with three lines, as shown in the figure.

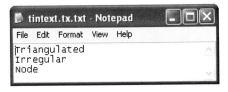

The `writelines` method can be used to modify the string for a particular line. For example:

```
>>> f = open("C:/Data/tintext.txt")
>>> lines = f.readlines()
>>> f.close()
>>> lines[2] = "Network"
>>> f = open("C:/Data/tintext.txt", "w")
>>> f.writelines(lines)
>>> f.close()
```

In this example, the `readlines` method is used to return the contents of the text file as a list. The file is then closed, and a new value is assigned to one of the elements in the list. The file is opened again in `write` mode and the `writelines` method is used to update the string value of that particular line. Running this code results in the text file, as shown in the figure.

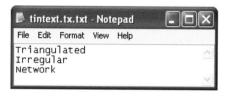

Keep in mind that new lines are not added automatically—you need to add them yourself using the line separator (\n). There is also no `writeline` method since the `write` method can be used for this.

Typically, you should close your files by calling the `close` method. When you open a file in `read` mode, this is not as critical because the file object is closed automatically when you quit the program using the file. However, closing your files is good practice because it prevents files from being needlessly locked. When writing to a file, you should always close it when you are finished, because Python may buffer in memory the data you have written, and if the program crashes, the data may not be written at all.

In the previous example, you saw how to write a new value to a particular line. Modifying a text file using Python is a commonly used technique. Many data files from other applications are in text format but are not directly usable by ArcGIS. Rather than manually manipulating the text file, you can use Python to automate this task. In the previous example, a change was made at a very specific location. It is much more common, however, to perform a find-and-replace search that will make changes to the text wherever necessary.

Consider the following example of a text file that contains geographic coordinates:

```
ID: 1, Latitude: 35.099722, Longitude: -106.527550
ID: 2, Latitude: 35.133015, Longitude: -106.583581
ID: 3, Latitude: 35.137142, Longitude: -106.650632
ID: 4, Latitude: 35.093650, Longitude: -106.573590
```

To make the file easier to use, you want it to look as follows:

```
1 35.099722 -106.527550
2 35.133015 -106.583581
3 35.137142 -106.650632
4 35.093650 -106.573590
```

One approach is to open the file in a text editor and perform a find-and-replace search. However, the same can be accomplished in Python. It requires that you iterate over the contents of a file, repeatedly performing the same action. There are several ways to iterate. In most cases, you can stick with one approach, but it is useful to know some of the different options, especially when trying to understand someone else's code.

One of the most basic ways of iterating over the contents of a file is to use the `read` method in a `while` loop. For example, the following code loops over every character in the file:

```
f = open("C:/Data/mytext.txt")
char = f.read(1)
while char:
    <function>
    char = f.read(1)
f.close()
```

The line `<function>` represents some procedure that represents the processing of each character. When you reach the end of the file, the `read` method returns an empty string and `char` is `False`, which ends the `while` loop.

Another way to write the same iteration is as follows:

```
f = open("C:/Data/mytext.txt")
while True:
    char = f.read(1)
    if not char: break
    <function>
f.close()
```

Although iterating over individual characters can be useful, it is more common to iterate over lines in a text file. There are a number of ways to accomplish this.

First, files in Python are *iterable*, which means you can use them directly in a `for` loop to iterate over their lines. The code is as follows:

```
f = open("C:/Data/mytext.txt")
for line in f:
    <function>
f.close()
```

This is perhaps the most elegant version of iteration over lines of text because the code is very short and makes direct use of the iterable nature of files.

Iteration can also be accomplished using the `readline` method. For example:

```
f = open("C:/Data/mytext.txt")
while True:
    line = f.readline()
    if not line: break
    <function>
f.close()
```

For relatively small files, you can also read the entire file in step using the `read` method (to read the entire file as a string) or the `readlines` method (to read the file into a list of strings). For example:

```
f = open("C:/Data/mytext.txt")
for line in f.readlines():
    <function>
f.close()
```

Reading the entire file using the `read` method or the `readlines` method can use excessive memory. One alternative is to use a `while` loop with the `readline` method. A second alternative is to use the `fileinput` module instead of the `open` function. This module enables you to create an object that you can iterate over in a `for` loop. For example:

```
import fileinput
for line in fileinput.input("C:/Data/mytext.txt")
    <function>
```

The approach you use is largely a matter of preference.

Now that you have seen the general structure of iteration over lines, it is time to look at an actual example and revisit the text file coordinates.txt, which looks like the example in the figure.

```
coordinates.txt - Notepad
File  Edit  Format  View  Help
ID: 1, Latitude: 35.099722, Longitude: -106.527550
ID: 2, Latitude: 35.133015, Longitude: -106.583581
ID: 3, Latitude: 35.137142, Longitude: -106.650632
ID: 4, Latitude: 35.093650, Longitude: -106.573590
ID: 5, Latitude: 35.071334, Longitude: -106.705875
ID: 6, Latitude: 35.062954, Longitude: -106.627941
ID: 7, Latitude: 35.076698, Longitude: -106.658666
ID: 8, Latitude: 35.125671, Longitude: -106.601285
ID: 9, Latitude: 35.123812, Longitude: -106.561119
ID: 10, Latitude: 35.200106, Longitude: -106.708698
ID: 11, Latitude: 35.196102, Longitude: -106.730564
ID: 12, Latitude: 35.145754, Longitude: -106.518095
ID: 13, Latitude: 35.148999, Longitude: -106.509894
ID: 14, Latitude: 35.098853, Longitude: -106.700963
ID: 15, Latitude: 35.095873, Longitude: -106.713954
```

Suppose you want to reduce this to a text file that includes only the relevant attribute values separated by a space, without the names of the fields (ID, Latitude, Longitude). You can accomplish this by iterating over the lines and using the replace method for string variables. When manipulating text files, it is good practice to save your result to a new file. This way you won't lose any of your original data if the code does not perform exactly as expected.

The following script opens an existing file in read mode and creates a new output file in write mode. A for loop is used to iterate over the lines in the input file. In the block of code that follows, the replace method is used three times to remove specific strings from each line. The resulting string is written to the output file. The code is as follows:

```
input = open("C:/Data/coordinates.txt")
output = open("C:/Data/coordinates_clean.txt", "w")
for line in input:
    str = line.replace("ID: ", "")
    str = str.replace(", Latitude:", "")
    str = str.replace(", Longitude:", "")
    output.write(str)
input.close()
output.close()
```

Running this code results in a text file like the example in the figure. →

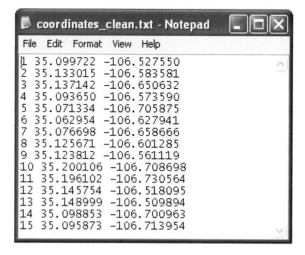

In chapter 8, you will see how a text file like this can be used to create geometries.

## Points to remember

- The ArcPy data access module, `arcpy.da`, provides support for editing and cursors.

- Cursors can be used to iterate over rows in a table. Iteration is typically accomplished using a `for` loop or a `with` statement.

- Search cursors are used to iterate over records. Insert cursors are used to add new records to a table. Update cursors are used to make changes.

- SQL query expressions can be used in Python using search cursors. Proper syntax is facilitated using the `AddFieldDelimiters` function.

- Table and field names can be validated using the `ValidateTableName` and `ValidateFieldName` functions, respectively. These functions convert all invalid characters into an underscore (_). The `CreateUniqueName` function can be used to create a unique name by adding a number to a name.

- Table and field names can be parsed into separate elements using the ArcPy parsing functions.

- The contents of text files can be manipulated in Python. The `open` function creates a file object and a number of methods can be used to read and write text, including `read`, `readline`, `readlines`, `write`, and `writelines`. One of the more common operations on files is to iterate over their contents performing the same manipulation repeatedly, such as replacing strings to make the text files more usable.

# Chapter 8
## Working with geometries

## 8.1 Introduction

This chapter describes how to work with geometries, including how to create geometry objects from existing feature classes and how to read the properties of these geometry objects. Individual features, such as points, polylines, and polygons, can be broken down into their vertices. Geometries can also be written by creating geometry objects from a list of coordinates. Being able to read and write geometries provides detailed control of feature classes, features, and the parts and vertices that make up features.

## 8.2 Working with geometry objects

Each feature in a feature class contains a set of points that define the vertices of the feature. These points can be accessed using geometry objects, such as Point, Polyline, PointGeometry, and MultiPoint, which returns an array of Point objects.

Accessing full geometry objects is somewhat time consuming. As a result, scripts that work with the full geometry objects of large datasets can become very slow. If you need only specific properties of the geometry, you can use geometry tokens as shortcuts to access geometry properties. For example, SHAPE@XY will return a tuple of x,y coordinates representing the feature's centroid, and SHAPE@LENGTH will return the feature's length as a double. On the other hand, SHAPE@ will return the full geometry object.

The following example will determine the combined length of all features in a polyline feature class:

```
import arcpy
fc = "C:/Data/roads.shp"
cursor = arcpy.da.SearchCursor(fc, ["SHAPE@LENGTH"])
length = 0
for row in cursor:
    length += row[0]
print length
```

## 8.3 Reading geometries

Each feature in a feature class consists of a set of points that define the vertices of point, polyline, or polygon features. In the case of a point feature class, each point consists of only a single vertex.

Polyline and polygon features consist of multiple vertices. Each vertex is a location defined by coordinates. The figure illustrates how points, polylines, and polygons are defined by vertices.

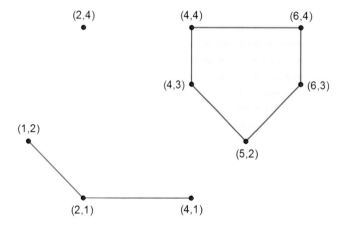

The points that define these vertices can be accessed using a search cursor. Point features return a single point object, and other feature types (polyline, polygon, multipoint) return an array of point objects. If a feature has multiple parts, an array containing multiple arrays of points is returned. Multipart features are revisited later in this chapter.

In the following example, a search cursor and a `for` loop are used to iterate over the rows of a point feature class called hospitals.shp. A geometry token is used to retrieve the x,y coordinates of the point objects, which are then printed. The script is as follows:

```
import arcpy
fc = "C:/Data/hospitals.shp"
cursor = arcpy.da.SearchCursor(fc, ["SHAPE@XY"])
for row in cursor:
    x, y = row[0]
    print("{0}, {1}".format(x,y))
```

The hospitals shapefile, which is a point feature class, is shown in the figure. Each point has a pair of x,y coordinates.

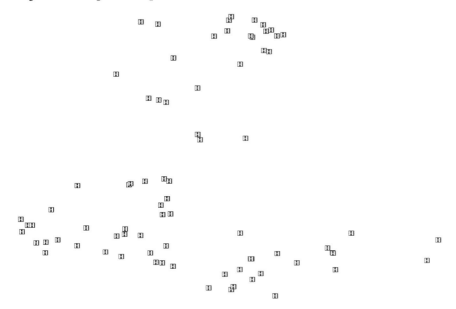

Running the script on the hospitals shapefile returns a list of x,y coordinate pairs, as follows:

```
535054.323998  1573193.51843
536331.986799  1568007.16872
537099.102795  1573622.42036
509801.659767  1577555.46667
516240.163452  1568947.90507
511539.047392  1576059.89036
513866.290182  1571538.02958
510005.314649  1574390.10495
516315.610359  1571701.11945
...
```

Point feature classes are relatively simple because there is only a single point object for each feature. For other types of feature classes such as polylines and polygons, an array of point objects is returned for each feature. To work with these arrays, an extra iteration is needed. In a typical script, a `for` loop is used to iterate over the rows in the table and a second `for` loop is used to iterate over the point objects in each array.

In the following example code, a for loop is used to iterate over the rows in a shapefile. For every row, the value of the OID (object identifier) field is printed—without it, you could not tell the start and end of each array of points. For each row, a geometry object is obtained, which consists of an array containing an array of point objects. The getPart method is used to obtain an array of point objects for the first (and only) part of the geometry. (**Note:** *Geometry parts are covered in more detail in the next section.*) A for loop is used to iterate over all the point objects in the array and print the x,y coordinates. The code is as follows:

```
import arcpy
from arcpy import env
env.workspace = "C:/Data"
fc = "roads.shp"
cursor = arcpy.da.SearchCursor(fc, ["OID@", "SHAPE@"])
for row in cursor:
    print("Feature {0}: ".format(row[0]))
    for point in row[1].getPart(0):
        print("{0}, {1}".format(point.X, point.Y))
```

The roads shapefile, which is shown in the figure, has vertices that are shown for emphasis—these are usually not visible for a polyline shapefile. The shapefile consists of three polylines, which have coincident geometry in one location.

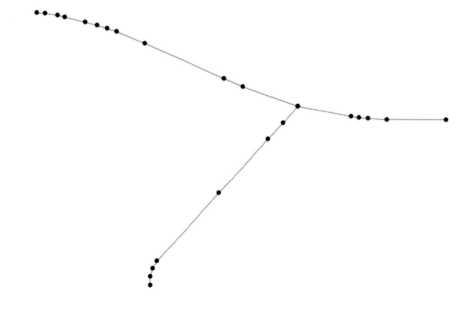

When the preceding script is run on the roads shapefile, the list of x,y
coordinates of the vertices in each of the three features is printed. The
coincident vertices representing the road intersection are highlighted here
for emphasis:

```
Feature 0:
531885.67639 1628468.19487
531516.812298  1628594.51778
531390.493653  1628643.4524
530857.141374  1628870.30004
530669.823179  1628944.1247
530605.296405  1628964.82938
530534.895627  1628986.13347
530458.155951  1629004.96119
530320.821908  1629036.68192
530272.489 1629046.89221
530191.019018  1629058.52374
530132.914148  1629065.05982
Feature 1:
532869.396535  1628383.24097
532476.944893  1628385.50147
532355.168561  1628393.21733
532296.13948  1628398.46043
532242.764587  1628403.90793
531885.67639 1628468.19487
Feature 2:
531885.67639 1628468.19487
531787.985312  1628360.42343
531688.961888  1628255.70514
531358.17318  1627909.01489
530942.021781  1627471.36419
530911.913574  1627422.25799
530897.352579  1627370.96183
530895.624892  1627312.96063
```

A couple of other things should be noted about this script. First, the
getPart method uses an index value of zero (0). This means the method
returns only the first part of the geometry object that has index value 0. For
regular (that is, single part) feature classes, the first part is also the only
part. If no index value is specified, the getPart method returns an array
containing an array of point objects. This is addressed in more detail in the
next section, on multipart features. Second, the script can be used for both
polyline and polygon feature classes.

# 8.4 Working with multipart features

Features in a feature class can have multiple parts, making them multipart features. Such features are sometimes needed when there are multiple physical parts to a feature but only one set of attributes. A classic example of a multipart feature is the state of Hawaii: each of the islands is its own part, but for Hawaii to be shown as a single record in the attribute table, these parts must form a single feature.

In the case of points, these features are referred to as multipoint, and in the case of polylines and polygons, they are referred to as multipart.

Whether a feature class is multipart can be determined using the shapeType property of the Describe function. Valid return values for this property are Point, Polyline, Polygon, MultiPoint, and MultiPatch, which is used to represent three-dimensional data. When a feature class is multipart, however, it does not mean that every feature in the feature class is multipart. The isMultipart property of the geometry object is used to determine whether a particular feature is multipart. The partCount property of the geometry object can be used to obtain the number of geometry parts of a feature.

The syntax for working with multipart geometries is very similar to the syntax for single-part features. The key difference is that for multipart features, an array containing multiple arrays of point objects is returned, instead of the single array of point objects for single-part features. A script working with geometry, therefore, has to iterate over not just the rows of a table (and over the array of point objects for polylines and polygons), but also over the array of parts for each geometry object.

The following example code illustrates how this is accomplished for polyline and polygon feature classes:

```
import arcpy
from arcpy import env
env.workspace = "C:/Data"
fc = "roads.shp"
cursor = arcpy.SearchCursor(fc, ["OID@", "SHAPE@"])
for row in cursor:
    print("Feature {0}: ".format(row[0]))
    partnum = 0
    for part in row[1]:
        print("Part {0}:".format(partnum))
        for point in part:
            print("{0}, {1}".format(point.X, point.Y))
        partnum += 1
```

This script works for both single-part and multipart features. For single-part features, the number of parts is one (1) and the code block in the `for` loop is run only once for each geometry object.

This script works for polyline and polygon feature classes, as well as for single-part and multipart features.

When applied to the same roads shapefile as before, the output is as follows:

```
Feature 0:
Part 0:
531885.67639 1628468.19487
531516.812298 1628594.51778
...
Feature 1:
Part 0:
532869.396535 1628383.24097
532476.944893 1628385.50147
...
Feature 2:
Part 0:
531885.67639 1628468.19487
531787.985312 1628360.42343
...
```

The script prints the feature and part number (starting with index 0) followed by the vertices in each part. For a single-part feature class, the part number will always be zero (0). When the script is run on multipart features, however, the vertices in each part are printed separately. For example, when the script is applied to a feature class of the state of Hawaii where all the islands are part of a single multipart feature, the output is as follows:

```
Feature 0:
Part 0:
893854.841767 2134701.2927
893846.673616 2134665.30299
...
Part 1:
912254.26006 2186306.54325
912224.615223 2186294.09826
...
Part 2:
912042.643477 2186622.39331
912012.925579 2186604.06848
...
Part 3:
822533.490168 2198560.73868
822527.657774 2198547.11324
...
Part 4:
829031.967738 2244295.66214
829042.696727 2244292.43166
...
```

**>>> TIP**

A single feature can have a very large number of vertices, so be careful when running scripts on large feature classes — printing all the vertices may take a very long time.

# 8.5 **Working with polygons with holes**

If a polygon contains holes, it will consist of a number of rings: one exterior ring and one or more interior rings. A ring is a closed path that defines a two-dimensional area. A path is a series of vertices with a starting vertex (from) and an ending vertex (to). A valid ring consists of a valid path in which the from and the to points of the ring have the same x,y coordinates. An exterior ring is defined as a clockwise ring and an interior ring is defined as a counterclockwise ring.

For a polygon with holes, the geometry object returns an array of point objects that contains the points of the exterior ring and all the inner rings. The exterior ring is always returned first, followed by the inner rings. A null point object—that is, a point object without values—is used as the separator between rings.

A script to read the geometry of polygons with holes is very similar to the script in the preceding section for multipart features. The third `for` loop is replaced by the following:

```
for point in part:
    if point:
        print("{0}, {1}".format(point.Y, point.Y))
    else:
        print "Interior Ring"
partnum += 1
```

The only addition consists of an `if-else` statement. For a geometry object with rings, null points are used as separators between rings. Therefore, if the next object following a null point is a point object, it means there is an interior ring. This block of code is run for every null point separator, resulting in the vertices of each interior ring being listed separately.

Polygons with holes are quite common, especially in feature classes that describe natural features, such as vegetation and soils. The example in the figure illustrates a typical soil polygon.

Running the script on the soil polygon results in the following output. There is only a single feature (Feature 0) that has an outer ring (Part 0) and multiple inner rings. In this example, the rings are not numbered. The output is as follows:

```
Feature 0:
Part 0:
549563.387926 1623728.65919
549615.387903 1623700.65924
...
Interior Ring
547042.700688 1625044.65659
546906.386611 1624984.65647
...
Interior Ring
548234.35805 1621938.65839
548210.169534 1621866.65838
...
Interior Ring
546255.274993 1623827.6562
546215.180853 1623815.65616
...
...
...
```

# 8.6 **Writing geometries**

New features can be created or updated using the insert and update cursors. A script can define a feature by creating point objects, populating their properties, and placing the point objects in an array. This new array can then be used to set the geometry of a feature.

For example, the following text file lists 21 points, each starting with an ID number and followed by an x-coordinate and a y-coordinate, with the coordinates of each pair separated by a space (" "). The coordinates of these points will be used to create a new polygon—notice that the coordinates of the first point and the last point in the list are identical. These coordinates are stored in a text (.txt) file called `points.txt`. The text file reads as follows:

```
1  542935  1619969
2  543015  1619964
3  543079  1619924
4  543095  1619896
5  543107  1619816
6  543099  1619768
7  543067  1619669
8  543047  1619629
9  543007  1619593
10  542979  1619577
11  542923  1619569
12  542883  1619577
13  542810  1619625
14  542738  1619649
15  542698  1619701
16  542690  1619733
17  542699  1619773
18  542719  1619821
19  542775  1619893
20  542883  1619953
21  542935  1619969
```

The `CreateFeatureclass` function can be used to create a new, empty feature class, which will be used to hold the new point objects whose coordinates are taken from the preceding list. The syntax of this tool is as follows:

```
CreateFeatureclass_management(out_path, out_name, {geometry_type}, ➡
➡ {template}, {has_m}, {has_z}, {spatial_reference}, {config_keyword}, ➡
➡ {spatial_grid_1}, {spatial_grid_2}, {spatial_grid_3})
```

The only required parameters are the path for the location of the new feature class (folder or geodatabase) and the name of the new feature class. The default value of the geometry is Polygon. There is no default for the spatial reference, so if none is specified, the coordinate system will be "unknown." The first part of the script is as follows:

```
import arcpy, fileinput, string
from arcpy import env
env.overwriteOutput = True
infile = "C:/Data/points.txt"
fc = "C:/Data/newpoly.shp"
arcpy.CreateFeatureclass_management("C:/Data", fc, "Polygon")
```

So far, this has created a new, empty feature class called `newpoly.shp`.

The point objects representing the vertices of the polygon can be created using the ArcPy `Point` class. These point objects have to be stored in an array. An array object can be created using the ArcPy `Array` class. In general, an array can contain any number of geoprocessing objects such as points, geometries, or spatial references. In this case, the array will contain point objects. In addition, an insert cursor is created to make it possible to create new rows—that is, new features. These lines of code are as follows:

```
cursor = arcpy.da.InsertCursor(fc, ["SHAPE@"])
array = arcpy.Array()
point = arcpy.Point()
```

Next, the properties of the point objects have to be set using the values in the text file. This requires the `fileinput` Python module to read the text file, and the `split` method to parse the text into separate strings for the point ID number, the x-coordinate, and the y-coordinate. These lines of code are as follows:

```
for line in fileinput.input(infile):
    point.ID, point.X, point.Y = line.split()
```

The `split` method returns a list of strings using the argument of the method as the delimiter. When no argument is specified (as is the case here), the `split` method uses consecutive whitespace as the delimiter. For each line, the `split` method returns a list of three strings. These values are then assigned to ID, X, and Y.

Finally, the script needs to iterate over the lines of the input text file and create a point object for every line. The result is a single array with 21 point objects. The completed script is as follows:

```
import arcpy, fileinput, os
from arcpy import env
env.workspace = "C:/Data"
infile = "C:/Data/points.txt"
fc = "newpoly.shp"
arcpy.CreateFeatureclass_management("C:/Data", fc, "Polygon")
cursor = arcpy.da.InsertCursor(fc, ["SHAPE@"])
array = arcpy.Array()
point = arcpy.Point()
for line in fileinput.input(infile):
    point.ID, point.X, point.Y = line.split()
    line_array.add(point)
polygon = arcpy.Polygon(array)
cursor.insertRow([polygon])
fileinput.close()
del cur
```

The result of the script is a new shapefile called `newpoly.shp` with a single polygon feature, as shown in the figure.

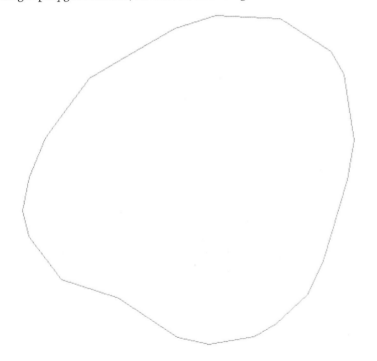

The example script is still relatively simple because it created only a single polygon with no other attributes. However, it illustrates the concept of using the `Point` and `Array` classes to create new geometry objects.

# 8.7 **Using cursors to set the spatial reference**

The spatial reference for a feature class describes the coordinate system, the spatial domain, and the precision. The spatial reference is typically set when the feature class is created. However, since specifying a spatial reference is not required, it results in an unknown coordinate system when none is specified. In this case, the Define Projection tool can be used to record the coordinate system information for the feature class.

A spatial reference applies to all the features in a feature class. By default, the spatial reference of the geometry of an object returned from a cursor is, therefore, the same as the spatial reference of the feature class opened by the cursor. In certain circumstances, however, you may be working with geometries that have a different spatial reference from the feature class—for example, if you have a feature class in a state plane coordinate system and you want to insert new features using a text file that has universal transverse Mercator (UTM) coordinates. In this case, you could set the spatial reference on the update or insert cursor to ensure proper conversions. You would open an insert cursor on the feature class and set the spatial reference of the cursor to UTM, thus declaring that the geometries to be inserted need to be converted from UTM to state plane.

You can also set the spatial reference of a search cursor. Specifying a spatial reference that is different from the spatial reference of the feature class results in geometries that are converted into the spatial reference of the cursor.

Consider the example of using a point feature class in state plane coordinates and writing a script that exports the x,y coordinate pairs of the point objects in decimal degrees. The SearchCursor function is used to establish a read-only cursor on the state plane coordinates of the feature class, but the spatial reference of this cursor is set to the desired geographic coordinate system, in decimal degrees. This is accomplished using the following code:

```
import arcpy
fc = "C:/Data/hospitals.shp"
prjfile = "C:/projections/GCS_NAD_1983.prj"
spatialref = arcpy.SpatialReference(prjfile)
cursor = arcpy.da.SearchCursor(fc, ["SHAPE@"], "", spatialref)
```

Next, an output file is created, using the open function. This opens the file in writing mode ("w") so that new lines of text can be written to it, as follows:

```
output = open("result.txt", "w")
```

The next step is to iterate over the rows, create a geometry object for each row, and write the x,y coordinates to the output file using the write method. This part of the code is as follows:

```
for row in cursor:
    point = row[0]
    output.write(str(point.X) + " " + str(point.Y) + "\n")
```

The coordinates are written as decimal degrees in a string, with a space (" ") separating the coordinates of each pair, and with a line break ("\n") for each point object. The very last step is to close the output file using the close method. The complete code is as follows:

```
import arcpy
from arcpy import env
env.workspace = "C:/Data"
fc = "hospitals.shp"
prjfile = "C:/Projections/GCS_NAD_1983.prj"
spatialref = arcpy.SpatialReference(prjfile)
cursor = arcpy.da.SearchCursor(fc, ["SHAPE@"], "", spatialref)
output = open("result.txt", "w")
for row in cursor:
    point = row[0]
    output.write(str(point.X) + " " + str(point.Y) + "\n")
output.close()
```

In this example, the spatial reference is set by using an existing projection file (.prj), but it can also be obtained from an existing feature class.

> *Note: Although setting the spatial reference on the search cursor can be used to convert geometries from one coordinate system to another, it is not very robust. In particular, setting any necessary datum transformation is not automatically taken into account and would need to be set in the script as part of the environment settings.*

# 8.8 Using geometry objects to work with geoprocessing tools

Inputs for geoprocessing tools often consist of feature classes. Sometimes, however, these feature classes do not yet exist and need to be created from geometry information. In this case, you can create a new feature class, populate the feature class using cursors, and then use the feature class in geoprocessing tools. This can become cumbersome, however. As an alternative, geometry objects can be used instead of both input and output feature classes to make geoprocessing simpler.

For example, the following code creates a list of geometry objects from a list of coordinates, and then uses the geometry objects as input to the Buffer tool, as follows:

```
import arcpy
from arcpy import env
env.workspace = "C:/Data"
coordlist = [[17.0, 20.0], [125.0, 32.0], [4.0, 87.0]]
pointlist = []
for x, y in coordlist:
    point = arcpy.Point(x,y)
    pointgeometry = arcpy.PointGeometry(point)
    pointlist.append(pointgeometry)
arcpy.Buffer_analysis(pointlist, "buffer.shp", "10 METERS")
```

In the example code, the geometry objects are created as a list of point objects. First, an empty list is created using `pointlist = []`. In the `for` loop, the list of coordinate pairs is used to create point objects using the `Point` class. These point objects are then used by the `PointGeometry` class to create geometry objects, which are appended to the list. The list becomes the input for the Buffer tool. An alternative would be to first create a feature class based on the list of coordinates, but if this feature class is not necessary for anything else, the use of geometry objects will result in more efficient code.

Geometry objects can also be created directly as the output of geoprocessing tools. For example, the following code uses an empty geometry object as the output of the Copy Features tool, and the result is a list of geometry objects, as follows:

```
import arcpy
fc = "C:/Data/roads.shp"
geolist = arcpy.CopyFeatures_management(fc, arcpy.Geometry())
length = 0
for geometry in geolist:
    length += geometry.length
print "Total length: " + length
```

The use of geometry objects can improve the efficiency of your code because it allows you to avoid the steps of having to create temporary feature classes and use a cursor to read through all the features.

## Points to remember

- The geometry object provides access to a number of properties, including length and area. Geometry tokens can be used as shortcuts to specific geometry properties.

- Individual vertices of geometry objects are stored as an array of point objects (or an array containing multiple arrays of point objects in the case of multipart features).

- New features can be created or updated using the insert and update cursors. A script can define a feature by creating point objects, populating their properties, and placing the point objects in an array. This new array can then be used to set the geometry of a feature.

- The spatial reference can be set on cursors to work with geometries in a coordinate system that is different from that of the feature class.

- Geometry objects can be used instead of feature classes as inputs and outputs for geoprocessing tools to make scripting easier.

# Chapter 9
## Working with rasters

## 9.1 Introduction

Rasters present a unique type of spatial data, and a number of geoprocessing tools are designed specifically to take advantage of the raster data structure. This chapter illustrates how to use regular ArcPy functions to list and describe rasters. ArcPy also includes a Spatial Analyst module referred to as `arcpy.sa`, which fully integrates map algebra into the Python environment, making scripting much more efficient. Map algebra operators are described, as well as functions and classes of the `arcpy.sa` module.

## 9.2 Listing rasters

The `ListRasters` function returns a Python list of rasters in a workspace. The syntax of the function is

```
ListRasters({wild_card}, {raster_type})
```

An optional `wild_card` parameter can be used to limit the list based on the name of the rasters. The optional `raster_type` parameter can be used to limit the list based on the type of raster—for example, JPEG or TIFF.

The following code illustrates the use of the `ListRasters` function to print a list of the rasters in a workspace:

```
import arcpy
from arcpy import env
env.workspace = "C:/raster"
rasterlist = arcpy.ListRasters()
for raster in rasterlist:
    print raster
```

The output would look something like the following:

```
elevation
landcover.tif
tm.img
```

The name of each raster is printed to the Interactive Window in PythonWin or the next line in the Python window, along with an optional file extension. For example, it is .img for the ERDAS IMAGINE format, .tif for the TIFF format, .jpg for the JPEG format, and so on. No file extensions are added for the Esri GRID (global resource information database) format or for rasters stored inside a geodatabase. Therefore, when no file extension is present, be sure to determine whether you are working with a GRID or with a raster dataset inside a geodatabase.

The parameters of the `ListRasters` function can be used to filter the results. For example, the following code prints a list of the rasters in the workspace that are in the ERDAS IMAGINE format:

```
import arcpy
from arcpy import env
env.workspace = "C:/raster"
rasterlist = arcpy.ListRasters("*", "IMG")
for raster in rasterlist:
    print raster
```

Once the names of the rasters are obtained, other functions can be used, including functions to describe the data as discussed in the next section.

## 9.3 Describing raster properties

Rasters can be described using the generic `Describe` function as already discussed in chapter 6. The `Describe` function returns the properties for a specified data element. These properties are dynamic, which means the properties that are present depend on the data type being described. For example, when the `Describe` function is used on rasters, a generic set of properties is present in addition to specific properties that are unique to the specific raster element.

Three different raster data elements can be distinguished:

1.  Raster dataset—a raster spatial data model that is stored on disk or in a geodatabase. Raster datasets can be stored in many formats, including TIFF, JPEG, IMAGINE, Esri GRID, and MrSID. Raster datasets can be single band or multiband.

2.  Raster band—one layer in a raster dataset that represents data values for a specific range in the electromagnetic spectrum or other values derived by manipulating the original image bands. Many types of satellite images, for example, contain multiple bands.

3.  Raster catalog—a collection of raster datasets defined in a table of any format, in which the records define the individual raster datasets that are included in the catalog. Raster catalogs can be used to display adjacent or overlapping raster datasets without having to combine them into a mosaic in one large file.

Properties for each of these elements vary. For example, the format (TIFF, JPEG, and others) is a property of the raster dataset and the cell size is a property of the raster band. The general `dataType` property can be used to determine the type of data element. All properties, however, are accessed using the same `Describe` function.

The following code illustrates the use of the `Describe` function, which returns an object with properties that can be accessed, in this case for printing:

```
import arcpy
from arcpy import env
env.workspace = "C:/raster"
raster = "landcover.tif"
desc = arcpy.Describe(raster)
print desc.dataType
```

For this example of a raster in TIFF format, the `dataType` property returns the type `RasterDataset`. Properties that are specific to raster datasets only include the following:

* `bandCount`—the number of bands in the raster dataset

* `compressionType`—the compression type (LZ77, JPEG, JPEG2000, or None)

* `format`—the raster format (GRID, IMAGINE, TIFF, and more)

* `permanent`—indicates the permanent state of the raster: `False` if the raster is temporary, `True` if the raster is permanent

* `sensorType`—the sensor type used to capture the image

Once it has been determined that an element is a raster dataset, these properties can be accessed. For example, the following code includes additional properties used to describe the TIFF file:

```
import arcpy
from arcpy import env
env.workspace = "C:/raster"
raster = "landcover.tif"
desc = arcpy.Describe(raster)
print desc.dataType
print desc.bandCount
print desc.compressionType
```

This particular .tif file is a single-band uncompressed TIFF, and therefore the property `bandCount` returns a value of `1` and `compressionType` returns a value of `None`.

Many other properties that are commonly associated with rasters can be accessed for individual raster bands only. For example, the cell resolution is a very important raster property, but individual bands within one raster dataset can have different resolutions. A number of properties are specific to raster bands, including the following:

- `height`—the number of rows

- `isInteger`—indicates whether the raster band is an integer type

- `meanCellHeight`—the cell size in y direction

- `meanCellWidth`—the cell size in x direction

- `noDataValue`—the NoData value of the raster band

- `pixelType`—the pixel type, such as 8-bit integer, 16-bit integer, single precision floating point, and others

- `primaryField`—the index of the field

- `tableType`—the class name of the table

- `width`—the number of columns

For single-band raster datasets, the band itself does not have to be specified (there is only one, after all) and the properties can be accessed directly by describing the raster dataset. For example, the following code determines the cell size and pixel type of a raster:

```
import arcpy
from arcpy import env
env.workspace = "C:/raster"
rasterband = "landcover.tif"
desc = arcpy.Describe(raster)
print desc.meanCellHeight
print desc.meanCellWidth
print desc.pixelType
```

For this particular example, the code returns values of 30.0 by 30.0 and U8—this means the cell size is 30 by 30 meters and the pixel type is an unsigned 8-bit integer. These properties do not report the unit type, which has to be obtained from the Spatial Reference property. For example, the following code determines the name of the spatial reference and the unit:

```
spatialref = desc.spatialReference
print spatialref.name
print spatialref.linearUnitName
```

For multiband rasters, however, the specific band needs to be specified. Without a particular band being specified, properties such as cell size, height, width, and pixel type cannot be accessed. Specific bands are referenced using Band_1, Band_2, and so on. The following code illustrates how the properties for a band in a multiband raster dataset are accessed:

```
import arcpy
from arcpy import env
env.workspace = "C:/raster"
rasterband = "img.tif/Band_1"
desc = arcpy.Describe(rasterband)
print desc.meanCellHeight
print desc.meanCellWidth
print desc.pixelType
```

*Note: Individual bands in a raster are sometimes referenced, for example, as Layer_1 and Layer_2, instead of Band_1 and Band_2.*

# 9.4 **Working with raster objects**

ArcPy also contains a Raster class that is used to reference a raster dataset. A raster object can be created in two ways: (1) by referencing an existing raster on disk and (2) by using a map algebra statement. The syntax for the Raster class is

```
Raster(inRaster)
```

The following code illustrates how to create a raster object by referencing a raster on disk:

```
import arcpy
myraster = arcpy.Raster("C:/raster/elevation")
```

When using map algebra statements, the code looks something like the following:

```
import arcpy
outraster = arcpy.sa.Slope("C:/raster/elevation")
```

In both cases, the resulting raster object can be used in Python statements and additional map algebra expressions. Raster objects have many properties, which are largely similar to those already discussed earlier in this chapter, including bandCount, compressionType, format, height, width, meanCellHeight, meanCellWidth, pixelType, spatialReference, and others. Similar to the Describe function, these properties are mostly read-only.

Raster objects have only one method: save. The raster object (the variable and associated dataset) returned from a map algebra expression is temporary by default. This means the variable and the referenced dataset are deleted when the variable goes out of scope—for example, when ArcGIS is closed or when a stand-alone script is closed. The save method can be called to make the raster object permanent. The syntax of the save method is

```
save({name})
```

In the earlier example, the raster object outraster is temporary but can be made permanent using the following code:

```
import arcpy
outraster = arcpy.sa.Slope("C:/raster/elevation")
outraster.save("C:/raster/slope")
```

It may appear somewhat counterintuitive that map algebra expressions result in temporary outputs. Keep in mind that a typical workflow using rasters can involve numerous steps. If only the final output is actually needed, keeping temporary outputs as intermediate steps results in fewer output files and lower storage requirements.

# 9.5 Working with the ArcPy Spatial Analyst module

ArcPy includes a Spatial Analyst module, `arcpy.sa`, to carry out map algebra and other operations. The functionality provided by the Spatial Analyst module is largely the same as that of the tools in the Spatial Analyst toolbox. For example, you can run the Slope tool by referencing the Slope tool in the Spatial Analyst toolbox or by importing the `arcpy.sa` module and directly referencing the Slope tool.

The Spatial Analyst module integrates map algebra into the Python environment. This is similar to the use of map algebra in such ArcToolbox geoprocessing tools as Raster Calculator, Single Output Map Algebra, and Multiple Output Map Algebra in earlier versions of ArcGIS. The ArcPy Spatial Analyst module has a series of operators to support map algebra operations.

The Spatial Analyst module provides access to all the raster geoprocessing tools in the Spatial Analyst toolbox. It offers an alternative way to run these tools that can be more efficient than running them using the Spatial Analyst toolbox. Consider the following code that runs the Slope tool:

```
import arcpy
elevraster = arcpy.Raster("C:/raster/elevation")
outraster = arcpy.sa.Slope(elevraster)
```

Notice that the Slope tool is called using `arcpy.sa.Slope`, which appears to follow the regular syntax used for all tools: `arcpy.<toolboxalias>.<toolname>`. However, the alternative `arcpy.<toolname>_<toolboxalias>` syntax does not apply here, and `arcpy.Slope_sa` is not valid. Because `sa` is a module, and not just the alias of a toolbox, the code can be simplified as follows:

```
import arcpy
from arcpy.sa import *
elevraster = arcpy.Raster("C:/raster/elevation")
outraster = Slope(elevraster)
```

The statement `from arcpy.sa import *` imports all the functions from the `arcpy.sa` module, and tools can therefore be called directly—for example, `Slope` versus `arcpy.sa.Slope`. Initially, this may not appear to be much of a saving, but imagine having several dozen raster functions in a single map algebra expression—omitting `arcpy.sa` several dozen times makes your code shorter and easier to read.

# 9.6 Using map algebra operators

In addition to providing access to all the Spatial Analyst geoprocessing tools, the `arcpy.sa` module also includes a number of map algebra operators. Most of these operators are available as geoprocessing tools under the Math toolset in the Spatial Analyst toolbox yet are also available as operators in Python. Consider the following example, which converts elevation values from feet to meters using the Times tool:

```
import arcpy
from arcpy.sa import *
elevraster = arcpy.Raster("C:/raster/elevation")
outraster = Times(elevraster, "0.3048")
outraster.save("C:/raster/elev_m")
```

Instead of using the Times tool, the map algebra operator (*) can be used. The second-to-last line of code would look as follows:

```
outraster = elevraster * 0.3048
```

This alternative is a bit shorter, but more importantly, it makes it possible to write elaborate map algebra expressions relatively easily.

Consider the example of a suitability model in which you create three different rasters, each representing a different factor in the suitability model. In the final analysis step, you want to add these three rasters together and determine the average suitability score. Your code could look something like the following:

```
import arcpy
from arcpy.sa import *
f1 = arcpy.Raster("C:/raster/factor1")
f2 = arcpy.Raster("C:/raster/factor2")
f3 = arcpy.Raster("C:/raster/factor3")
temp1raster = Plus(f1, f2)
temp2raster = Plus(temp1raster, f3)
outraster = Divide(temp2raster, "3")
outraster.save("C:/raster/final")
```

The Plus tool has to be used twice to add all three rasters together because the tool can use only two inputs at a time. The Divide tool is used to divide the sum of the three rasters by 3. Using map algebra expressions, this code can be reduced as follows:

```
import arcpy
from arcpy.sa import *
f1raster = arcpy.Raster("C:/raster/factor1")
f2raster = arcpy.Raster("C:/raster/factor2")
f3raster = arcpy.Raster("C:/raster/factor3")
outraster = (f1 + f2 + f3) / 3
outraster.save("C:/raster/final")
```

This saves a fair amount of code and also makes it easier to recognize the map algebra expression.

If this kind of statement looks familiar, it is probably because you have used Raster Calculator before. In earlier versions of ArcGIS, it was a tool from the Spatial Analyst toolbar, but in version 10, it was added as a tool in the Spatial Analyst toolbox.

The Raster Calculator tool dialog box looks like the example in the figure. ➤

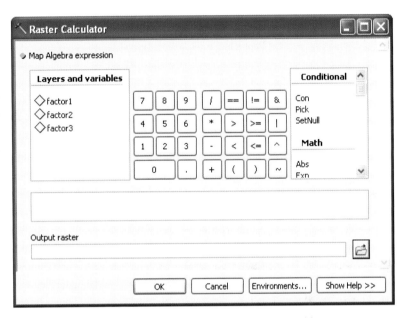

The dialog box allows you to create an expression by selecting from the layers and the tools. For example, the expression needed for the suitability model would look like the example in the figure. ➤

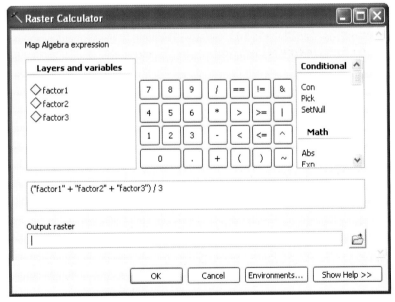

It looks very much like the Python code used earlier. In effect, the map algebra operators in the `arcpy.sa` module allow you to create Raster Calculator-style expressions directly in Python. You can also call the Raster Calculator tool using the following syntax:

```
RasterCalculator(expression, output_raster)
```

However, this will not make your code much more efficient unless you are using an expression that has map algebra operators that are not available directly in Python.

Table 9.1 shows a list of the map algebra operators that are available from the `arcpy.sa` module. They are grouped into four categories: Arithmetic, Bitwise, Boolean, and Relational.

**Table 9.1  Map algebra operators**

| Category | Operator | Description | Spatial Analyst tool |
|---|---|---|---|
| Arithmetic | - | Subtraction | Minus |
| | - | Unary Minus | Negate |
| | % | Modulo | Mod |
| | * | Multiplication | Times |
| | / | Division | Divide |
| | // | Integer Division | N/A |
| | + | Addition | Plus |
| | + | Unary Plus | N/A |
| | ** | Power | Power |
| Bitwise | >> | Bitwise Right Shift | Bitwise Right Shift |
| | << | Bitwise Left Shift | Bitwise Left Shift |
| Boolean | ~ | Boolean Complement | Boolean Not |
| | & | Boolean And | Boolean And |
| | ^ | Boolean Exclusive Or | Boolean XOr |
| | \| | Boolean Or | Boolean Or |
| Relational | < | Less Than | Less Than |
| | <= | Less Than or Equal To | Less Than Equal |
| | > | Greater Than | Greater Than |
| | >= | Greater Than or Equal To | Greater Than Equal |
| | == | Equal To | Equal To |
| | != | Not Equal To | Not Equal |

A detailed discussion of each operator is beyond the scope of this book, but some observations are in order:

- Most operators have an equivalent tool in the Math toolset, but two do not: // (integer division) and + (unary plus). However, the same tasks can be accomplished using a combination of tools.

- Many of the tools in the Math toolset do not have an equivalent map algebra operator in Python, including commonly used tools such as Abs, Int, Float, Exp10, Log10, and many others.

# 9.7 Using the `ApplyEnvironment` function

In addition to the geoprocessing tools in the Spatial Analyst toolbox, there is one more function: the `ApplyEnvironment` function. This function copies an existing raster and applies the current environment settings. The syntax of the function is

```
ApplyEnvironment(in_raster)
```

This function allows you to change things like the extent or the cell size or to apply an analysis mask. The following code illustrates how the `ApplyEnvironment` function is used to set a new cell size of 30 and apply an analysis mask based on an existing shapefile:

```
import arcpy
from arcpy import env
from arcpy.sa import *
elevfeet = arcpy.Raster("C:/raster/elevation")
elevmeter = elevfeet * 0.3048
env.cellsize = 30
env.mask = "C:/raster/studyarea.shp"
elevrasterclip = ApplyEnvironment(elevmeter)
elevrasterclip.save("C:/raster/dem30_m")
```

Not all environment settings apply to the `ApplyEnvironment` function. They are limited to the following: Cell Size, Current Workspace, Extent, Mask, Output Coordinate System, Scratch Workspace, and Snap Raster. These are the most relevant environment settings when working with rasters.

# 9.8 **Using classes of the** `arcpy.sa` **module**

The `arcpy.sa` module also contains a number of classes that are used for defining parameters of raster tools. Typically, these classes are used as shortcuts for tool parameters that would otherwise require a more complicated string value.

Consider the example of the Reclassification tool. With this tool, raster cells are given a new value based on a reclassification table. The tool dialog box in the figure shows an example of a land-use raster being reclassified into a number of values as part of a suitability model.

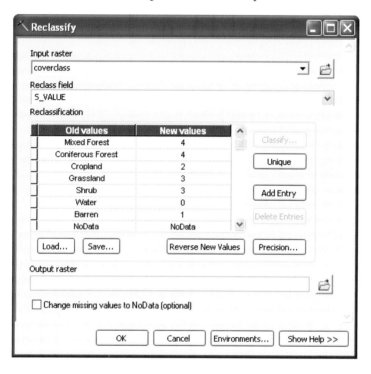

The syntax of the Reclassify tool is as follows:

```
Reclassify(in_raster, reclass_field, remap, {missing_values})
```

Typing all the values of this table would be rather complicated since this table can have many different entries. Instead, the `remap` parameter is expressed as a remap object. There are two different Remap classes, depending on the nature of the reclassification:

1.  `RemapValue`—a list of individual input values to be reclassified

2.  `RemapRange`—a list identifying ranges of input values to be reclassified

The syntax of the `RemapValue` object is

```
RemapValue(remapTable)
```

A `remapTable` object is defined using a Python list of lists that each contain old and new values, similar to the reclassification table on the tool dialog box. The syntax of a remap table for use in a `RemapValue` object is

```
[[oldValue1, newValue1], [oldValue2, newValue2], ...]
```

The following code illustrates the use of a remap object to carry out a reclassification of a raster representing land use:

```
import arcpy
from arcpy import env
from arcpy.sa import *
env.workspace = "C:/raster"
myremap = RemapValue([["Barren", 1], ["Mixed Forest", 4], ["Coniferous
Forest", 0], ["Cropland", 2], ["Grassland", 3], ["Shrub", 3], ["Water",
0]])
outreclass = Reclassify("landuse", "S_VALUE", myremap)
outreclass.save("C:/raster/lu_recl")
```

The `RemapRange` object works in a similar manner but uses value ranges rather than individual values. The syntax of a remap table for use in a `RemapRange` object is

```
[[startValue, endValue, newValue], ...]
```

The following code illustrates the use of a remap object to carry out a reclassification of a raster of elevation:

```
import arcpy
from arcpy import env
from arcpy.sa import *
env.workspace = ("C:/raster")
myremap = RemapRange([[1, 1000, 0], [1000, 2000, 1], [2000, 3000, 2],
[3000, 4000, 3]])
outreclass = Reclassify("elevation", "TYPE", myremap)
outreclass.save("C:/raster/elev_recl")
```

Notice that the end value of the first range is the same as the start value of the second range, and so on. This type of remap table is common when data is continuous, as in the case of a raster of elevation. In addition to the Reclassify tool, remap tables are also used in the Weighted Overlay tool.

There are many other classes in the `arcpy.sa` module. They can be grouped into a number of categories based on logical functionality. Table 9.2 lists these categories.

**Table 9.2  Categories of classes in the** `arcpy.sa` **module**

| Category | Description |
| --- | --- |
| Fuzzy Membership | Defines the membership function for fuzzy logic analysis |
| Horizontal Factor | Identifies the horizontal factor for the Path Distance tool |
| Kriging Models | Develops the model for creating surfaces with kriging |
| Neighborhood | Defines the input neighborhood for a series of tools |
| Overlay | Creates input tables for the Weighted Overlay and Weighted Sum tools |
| Radius | Identifies a radius for the IDW and Kriging tools |
| Remap | Defines various remap tables used in reclassification |
| Time | Identifies the time interval to use in the solar radiation tools |
| Topo Input | Defines the input to the Topo To Raster tool |
| Vertical Factor | Identifies the vertical factor for the Path Distance tool |

Among the more widely used classes, in addition to the Remap classes already discussed, are the Neighborhood classes, which define neighborhoods of different shapes and sizes. Consider, for example, the Focal Statistics tool. This tool, as well as other tools in the Neighborhood toolbox, requires the definition of a specific neighborhood.

The neighborhood settings vary with the type of neighborhood. For example, for the default rectangular neighborhood, settings include height and width in cell or map units. However, for the wedge neighborhood, the parameters include start angle and end angle and a radius in cell or map units.

Because of the variability of these parameters, neighborhood functions include a neighborhood object. For example, the syntax of the Focal Statistics tool is as follows:

```
FocalStatistics(in_raster, {neighborhood}, {statistics_type}, ➥
➥ {ignore_nodata})
```

There are six different neighborhood objects, each with its own unique set of parameters. They are as follows:

1. `NbrAnnulus`—defines an annulus, or ringlike, neighborhood, which is created by specifying an inner circle's and outer circle's radii in either map units or number of cells.

2. `NbrCircle`—defines a circular neighborhood, which is created by specifying the radius in either map units or number of cells.

3. `NbrIrregular`—defines an irregular neighborhood, which is created by a kernel file.

4. `NbrRectangle`—defines a rectangular neighborhood, which is created by specifying the height and the width in either map units or number of cells.

5. `NbrWedge`—defines a wedge neighborhood, which is created by specifying a radius and two angles in either map units or number of cells.

6. `NbrWeight`—defines a weight neighborhood, which identifies the cell positions to be included within the neighborhood and the weights for multiplying the cell values of the input raster.

For example, the syntax of the `NbrRectangle` object is

```
NbrRectangle({width}, {height}, {units})
```

The following code defines a neighborhood object and uses it in the `FocalStatistics` function:

```
import arcpy
from arcpy import env
from arcpy.sa import *
env.workspace = "C:/raster"
mynbr = NbrRectangle(5, 5, "CELL")
outraster = FocalStatistics("landuse", mynbr, "VARIETY")
outraster.save("C:/raster/lu_var")
```

In this example, the output is a raster of land cover based on a rectangular neighborhood of 5 cells by 5 cells.

## 9.9 Using raster functions to work with NumPy arrays

Two more raster functions need to be mentioned: `NumPyArrayToRaster` and `RasterToNumPyArray`. These are regular ArcPy functions, not functions of the `arcpy.sa` module. These two functions allow for conversions between rasters and NumPy arrays. A NumPy array is designed to work with very large arrays. NumPy itself is a package used for scientific computing with Python. Among other things, it provides a very powerful *n*-dimensional array object. This type of object makes it possible to move data between databases. For example, the SciPy package contains numerous algorithms that may be useful for a particular application, such as Fourier transforms, maximum entropy models, and multidimensional image processing. Rather than trying to build a tool in ArcGIS that carries out these specialized functions, you could write a script tool that converts a raster to a NumPy array, and then calls specialized functions from the SciPy package. A generic script would look as follows:

```
import arcpy, scipy
from arcpy.sa import *
inRaster = arcpy.Raster("C:/raster/myraster")
my_array = RasterToNumPyArray(inRaster)
outarray = scipy.<module>.<function>(my_array)
outraster = NumPyArrayToRaster(outarray)
outraster.save("C:/raster/result")
```

This is a simplified example and references a generic SciPy function, yet it illustrates how NumPy array functions can be used to export data for processing in another environment and to import the result back into an ArcGIS-compatible format—all within the same Python script. More information on NumPy (Numerical Python) and SciPy (Scientific Library for Python) can be found at http://numpy.scipy.org and http://www.scipy.org, respectively.

## Points to remember

- The `ListRasters` function is used to list rasters in a workspace. The `Describe` function is used to describe raster datasets and raster bands. Properties of objects returned by the `Describe` function are dynamic—that is, they depend on the nature of the data type.

- The `arcpy.sa` module integrates map algebra into the Python environment. In addition to providing access to all the tools in the Spatial Analyst toolbox, the `arcpy.sa` module contains a number of map algebra operators, which make scripting with rasters more efficient.

- The `arcpy.sa` module also contains a number of classes, which are primarily used for defining certain types of parameters of raster tools.

- Conversion functions are available to export a raster to a NumPy array, which makes it possible to use analysis functions from other Python libraries such as SciPy.

```
rt arcpy
port random
rom arcpy import env
env.overwriteoutput = T
inputfc = arcpy.GetPara
outputfc = arcpy.GetPar
utcount = int(arcpy.Ge
esc = arcpy.Describe(i
ist = []
mlist = []
= desc.OIDField
cpy.SearchCurs
next()

etValue(
d(id)
t()
) <
.o
d(
li
om
me
engt
ayer_

ures_mana
```

# Part 3
## Carrying out specialized tasks

# Chapter 10
## Map scripting

## 10.1 Introduction

This chapter describes the ArcPy mapping module, also referred to as `arcpy.mapping`. The ArcPy mapping module helps automate mapping tasks, including managing map documents, data frames, layer files, and the data within these files. There is also support for the automation of map export and printing, as well as the creation of PDF map books.

## 10.2 Working with the ArcPy mapping module

The ArcPy mapping module can be used to automate ArcMap workflows to speed up repetitive tasks. Some typical examples of uses for the ArcPy mapping module are as follows:

- Finding a layer with a particular data source and replacing it with another data source

- Modifying the display properties of a specific layer in multiple ArcMap documents

- Generating reports that describe the properties of ArcMap documents, including data sources, layers with broken data links, information on the spatial reference of data frames, and more

The highly visual ArcMap environment is the recommended application for creating new map documents and creating map layers and map layouts. Once they are created, however, the ArcPy mapping module can be used for scripting to automate certain mapping tasks, especially repetitive tasks across large numbers of map documents and elements. The ArcPy mapping module cannot be used to customize the ArcMap interface, but it allows you to automate many of the tasks you would normally carry out there.

Working with the ArcPy mapping module follows the workflow that would be used within an ArcMap session. For example, a typical workflow could consist of opening a map document, modifying properties of a data frame, adding a layer, modifying the properties of that layer, changing several elements of the page layout, and then exporting the map to a PDF file. These steps can be automated using scripts that employ the functions and classes of the ArcPy mapping module.

## 10.3 Opening map documents

A map document, or MXD, is stored as an .mxd file on disk—for example, C:\Mapping\Study_Areas.mxd. The ArcPy mapping module allows you to open and manipulate .mxd files, in addition to layer (.lyr) files, which contain properties for individual layers.

There are two ways to start working with a map document using the ArcPy mapping module: (1) use the map document from the current ArcMap session or (2) reference an existing .mxd file stored on disk. The `MapDocument` function is used to accomplish both. The syntax of the `MapDocument` function is

```
MapDocument(mxd_path)
```

The path is a string representing an .mxd file on disk. The following code opens a map document:

```
mapdoc = arcpy.mapping.MapDocument("C:/Mapping/Study_Areas.mxd")
```

To use the current map document in ArcMap, the keyword `CURRENT` (in all uppercase letters) is used:

```
mapdoc = arcpy.mapping.MapDocument("CURRENT")
```

To use the `CURRENT` keyword, ArcMap must be running because the `MapDocument` object will reference the map document currently loaded into ArcMap. Script tools that use the `CURRENT` keyword must be run from within ArcMap to run properly. Creating a script tool is covered in chapter 13. Background processing must be disabled to properly use the current map document. On creating a script tool, one of the properties that can be set is "Always run in foreground"—this is recommended when using the `CURRENT` keyword because it overrides the default background processing settings of the current ArcMap session.

When an existing .mxd file is used, the script can be run independently of ArcMap. The use of a system path to open an ArcMap document is recommended, however, because it makes the script more versatile and gives more control over how the script is run. Still, using the CURRENT keyword can be very useful for quickly testing code in the Python window.

The MapDocument object provides access to many different properties and methods of map documents. The MapDocument object also provides access to the other objects within a map document. The MapDocument object is a required parameter for many functions in the ArcPy mapping module. As a result, the MapDocument object is typically one of the first object references created in a mapping script.

Once the MapDocument object is created, properties of the map document can be modified. Before looking at how these changes are made, first consider how they are saved. If you are working with a map document in ArcMap and you make a change, such as adding a layer, there are two ways to save the .mxd file: Save and Save As. When Save is used, the changes are saved to the same .mxd file; when Save As is used, the changes are saved to a new .mxd file that you specify. In a scripting environment, however, the MapDocument variable always points to the original map document on disk or currently in memory. So there is no Save As option in the scripting environment and MapDocument uses only the save and saveACopy methods. However, saveACopy accomplishes the same thing as the Save As option in ArcMap, and also allows you to save the file to a previous version.

After changes are made to the current map document, the map may not automatically be updated with every line of code. The functions RefreshActiveView and RefreshTOC can be used to refresh the map document. This is similar to using the Refresh option in ArcMap (from the menu bar, click View > Refresh).

When a MapDocument object is referenced in a script, the .mxd file is locked. This prevents other applications from making changes to the file. It is therefore good practice to remove the reference to a map document when it is no longer needed in a script by using the Python del statement. A mapping script therefore often has a structure that looks something like the following:

```
import arcpy
mapdoc = arcpy.mapping.MapDocument("C:/Mapping/Study_Areas.mxd")
<code that modifies map document properties>
mapdoc.save()
del mapdoc
```

*Note:* When a script is finished running, Python automatically removes references to map documents, so the del statement is not required but still reduces the likelihood of unwanted locks.

# 10.4 Accessing map document properties and methods

The properties of a `MapDocument` object include most of the properties found on the Map Document Properties dialog box (from the ArcMap menu bar, click File > Map Document Properties). This includes properties such as the title and author of the map document, the date the map document was last saved, and whether the Relative paths check box has been selected. A complete description of these properties can be found in the ArcPy documentation in ArcGIS Desktop Help.

In addition to properties, the `MapDocument` object provides a number of methods. These include the `save` and `saveACopy` methods already mentioned, as well as methods for working with thumbnail images (`deleteThumbnail` and `makeThumbnail`), and methods for modifying workspaces (`findAndReplaceWorkspacePaths` and `replaceWorkspaces`). These last two methods are described in more detail in section 10.7 on fixing broken data sources.

In the following example, the `CURRENT` keyword is used to obtain the map document currently open in ArcMap, and the `filePath` property is used to print the system path for the .mxd file:

```
import arcpy
mapdoc = arcpy.mapping.MapDocument("CURRENT")
path = mapdoc.filePath
print path
del mapdoc
```

Running this code prints a system path. For example:

```
C:\Maps\final.mxd
```

The `del` statement ensures that the map document lock is removed.

The following example updates the current map document's title and saves the .mxd file:

```
import arcpy
mapdoc = arcpy.mapping.MapDocument("CURRENT")
mapdoc.title = "Final map of study areas"
mapdoc.save()
del mapdoc
```

As you review these basic examples, remember that they can be used to automate more complex tasks, such as making changes to multiple map documents rather than just the current one.

# 10.5 **Working with data frames**

Map documents contain one or more data frames and each data frame
typically contains one or more layers. Data frames and layers are
perfect objects for use in lists, which can help automate tasks. The
`ListDataFrames` function returns a list of `DataFrame` objects in a map
document. The syntax is as follows:

```
ListDataFrames(map_document, {wild_card})
```

Once you have a list of data frames in a map document, you can look
through them to examine or modify their properties. Running the following
code prints a list of all the data frames in a map document:

```
import arcpy
mapdoc = arcpy.mapping.MapDocument("CURRENT")
listdf = arcpy.mapping.ListDataFrames(mapdoc)
for df in listdf:
    print df.name
del mapdoc
```

If you want to work with just one of the data frames, you can use its index
number, as follows:

```
print listdf[0].name
```

The order of a list of data frames is the same as the order used in the
ArcMap table of contents.

    `DataFrame` object properties, such as map extent, scale, rotation, and
spatial reference, use map units. Other properties use page units to position and size the data frame on the layout page. Data frames are also used
to access other objects—for example, the `ListLayers` function is used to
access the layers in each data frame. You can then loop through the layers
to get their properties. It is therefore important to uniquely name the data
frames within a single map document.

There are quite a number of data frame properties, which are described in the ArcPy documentation in ArcGIS Desktop Help. The properties of the `DataFrame` object are a subset of all the properties on the Data Frame Properties dialog box in ArcMap (right-click a data frame in the Table Of Contents window and click Properties). ➔

Scripting does not provide access to all the properties on the Data Frame Properties dialog box, and conversely, some `DataFrame` object properties are not on the Data Frame Properties dialog box. For example, the scale of a data frame can be set using scripting, but in ArcMap, it is accomplished by using a tool on the Standard toolbar.

In most cases when you work with a map document, you are not interested in changing all the properties of a data frame, but only a few selected ones. For example, in the following code, the spatial reference of all data frames in a map document is set to the same spatial reference as that of a specific shapefile, and the scale of all data frames is set to 1:24,000:

```
import arcpy
dataset = "C:/map/boundary.shp"
spatialRef = arcpy.Describe(dataset).spatialReference
mapdoc = arcpy.mapping.MapDocument("C:/map/final.mxd")
for df in arcpy.mapping.ListDataFrames(mapdoc):
    df.spatialReference = spatialRef
    df.scale = 24000
del mapdoc
```

In addition to the properties already discussed, the `DataFrame` object also has two methods: `panToExtent` and `zoomToSelectedFeatures`. The `panToExtent` method maintains the data frame scale but pans and centers the data frame extent based on the properties of an `Extent` object, which has to be provided as a parameter. An `Extent` object is a rectangle specified by providing the coordinates of the lower-left corner and the upper-right corner in map units. In most cases, this property is derived from an existing object, such as a feature or a layer. For example, the `getExtent` method can be used to obtain the extent of a layer. The following code pans the extent of a data frame called `df` based on the extent of the features in a layer object called `lyr`:

```
df.panToExtent(lyr.getExtent())
```

The `zoomToSelectedFeatures` method is similar to the ArcMap operation Selection > Zoom to Selected Features. Running the following code zooms to the extent of all selected features in a data frame called `df`:

```
df.zoomToSelectedFeatures()
```

If no features are selected, the code will zoom to the full extent of all layers.

# 10.6 Working with layers

A data frame typically contains one or more layers and the `Layer` object is essential to managing these layers. The `Layer` object provides access to many different layer properties and methods. There are two ways to reference `Layer` objects. The first approach is to use the `Layer` function to reference a layer (.lyr) file on disk. It is similar to how map document files (.mxd) are referenced. The syntax of the `Layer` function is

```
Layer(lyr_file_path)
```

The parameter of the `Layer` function is the full path and file name of an existing .lyr file. For example:

```
Lyr = arcpy.mapping.Layer("C:/Mapping/study.lyr")
```

The second approach is to use the `ListLayers` function to reference the layers in an .mxd file, or just the layers in a particular data frame in a map document, or the layers within a .lyr file. The syntax of the `ListLayers` function is

```
ListLayers(map_document_or_layer, {wild_card}, {data_frame})
```

The only required element is a map document or layer file. An optional wild card can be used to limit the result. An optional data frame variable can be used that references a specific `DataFrame` object. For example, the following code returns a list of all the layers in an ArcMap document, and then prints the names of all the layers:

```
import arcpy
myDoc = arcpy.mapping.MapDocument("CURRENT")
lyrlist = arcpy.mapping.ListLayers(mapdoc)
for lyr in lyrlist:
    print lyr.name
```

To access just the layers in a specific data frame, the `DataFrame` object has to be referenced as a parameter. In the following example, the `ListLayers` function returns only the layers in the data frame that have index number 0. The `wild_card` parameter is skipped using an empty string (`""`).

```
import arcpy
myDoc = arcpy.mapping.MapDocument("CURRENT")
dflist = arcpy.mapping.ListDataFrames(mapdoc)
lyrlist = arcpy.mapping.ListLayers(mapdoc, "", dflist[0])
for lyr in lyrlist:
    print lyr.name
```

The following code illustrates how to reference the layers in a .lyr file on disk and print the names of the layer objects:

```
import arcpy
lyrfile = arcpy.mapping.Layer("C:/Data/mylayers.lyr")
lyrlist = arcpy.mapping.ListLayers(lyrfile)
for lyr in lyrlist:
    print lyr.name
```

Once you reference one or more `Layer` objects using either the `Layer` or `ListLayers` function, you have access to many of the common layer properties found on the Layer Properties dialog box in ArcMap. The `Layer` object also provides methods for saving layer files.

There are many types of layers in ArcMap and not all of them work in the same manner. Three of the layer categories are commonly used: feature layers, raster layers, and group layers. Properties of the `Layer` object can be used to identify the category you are working with and the `supports` method can be used to test the properties a layer supports. For example, a definition query would work only on a feature layer. But rather than remembering this aspect or checking it manually, you could use the `supports` method to test whether a particular layer supports a particular property.

*Note: Layers can get a bit confusing when working with a group layer. In this case, a single .lyr file can contain more than one layer, which is why the* `ListLayers` *function can have a .lyr file as a parameter. For a .lyr file with only a single layer, which is typical, the* `ListLayers` *function returns a list object with a single value. For the same .lyr file, however, the* `Layer` *function returns a single* `Layer` *object.*

There are a number of other more specialized layers, such as annotation subclasses, network datasets, topology datasets, and others. These layers may also require testing of properties to ensure they are supported.

`Layer` objects have a number of properties. These include the name of the layer, the name of the layer dataset, the ability to set a definition query, the ability to turn on the display of labels, and a number of display properties, such as brightness, contrast, and transparency. A complete description of all the properties of a `Layer` object can be found in the ArcPy documentation in ArcGIS Desktop Help. In the first version of ArcPy released with ArcGIS 10.0, emphasis was given to the properties that were most likely to benefit from automation. Additional properties are included with ArcGIS 10.1, such as layer symbology and access to a layer's time properties. Other properties may be included as well in future versions of the ArcPy mapping module.

A few examples will serve to illustrate the use of layer properties. For example, the following code turns on all the labels for the layers in the current map document using the `showLabels` property:

```
import arcpy
myDoc = arcpy.mapping.MapDocument("CURRENT")
dflist = arcpy.mapping.ListDataFrames(mapdoc)
lyrlist = arcpy.mapping.ListLayers(mapdoc, "", dflist[0])
for lyr in lyrlist:
    lyr.showLabels = True
del lyrlist
```

Instead of changing the properties of all the layers in a map document or a data frame, the layer properties can also be used to find a layer with a particular name. For example, the following code searches for a layer called "hospitals":

```
import arcpy
myDoc = arcpy.mapping.MapDocument("CURRENT")
lyrlist = arcpy.mapping.ListLayers(mapdoc)
for lyr in lyrlist:
    if lyr.name == "hospitals":
        lyr.showLabels = True
del lyrlist
```

Layer names can be a bit confusing. The name of a layer is what is shown in the ArcMap table of contents. This may or may not be the same as the name of the source dataset for the layer. In any case, the name of a layer does not have an extension. So the name of a feature class could be hospitals.shp, but as a layer in ArcMap, the name of the layer is hospitals.

Also remember that strings are case sensitive, so `Hospitals` is different from `hospitals`. To make your statements insensitive to case, you can use basic string operators. For example:

```
if lyr.name.lower() == "hospitals":
```

Several other layer properties involve names. The `datasetName` property returns the name of the layer dataset as it appears in the workspace. This does not, however, include any file extensions. The `dataSource` property returns the full path of the layer dataset. So for a layer that appears as Hospitals in the ArcMap table of contents, the `datasetName` property may be hospitals and the `dataSource` property may be C:\Data\hospitals.shp. Both `datasetName` and `dataSource` are read-only properties, whereas the name property is read/write and can be changed. Finally, there is the longName property, which is useful for describing group layers because it includes the group layer and sublayer names.

A number of methods exist for `Layer` objects. These include the `save` and `saveACopy` methods, which save a .lyr file. The `findAndReplaceWorkspacePath` and `replaceDataSource` methods are used to manipulate workspaces and are covered in more detail in the next section.

Because not all types of layers support the same properties, the `supports` method can be used to determine whether a layer supports a particular property. This makes it possible to test whether a layer supports a property before trying to get or set its value. This reduces the need for error checking. In the earlier example where the labels were shown for all layers in a data frame, it would make sense to first use the `supports` method to test whether this property is supported for each layer.

The syntax of the `supports` method is

```
supports(layer_property)
```

The parameter, in this case, would consist of one of the `Layer` object properties, such as `brightness`, `contrast`, `datasetName`, or others. The `supports` method returns a Boolean value, so the example code to test whether labeling is possible would look as follows:

```
import arcpy
myDoc = arcpy.mapping.MapDocument("CURRENT")
dflist = arcpy.mapping.ListDataFrames(mapdoc)
lyrlist = arcpy.mapping.ListLayers(mapdoc, "", dflist[0])
for lyr in lyrlist:
    if lyr.supports("SHOWLABELS") == True:
        lyr.showLabels = True
del lyrlist
```

If you are unsure whether a layer supports a particular property, use the `supports` method to test it. Otherwise, you will need to use an error-trapping method, such as a `try-except` statement, which is covered in chapter 11.

In addition to properties and methods of layer objects, there are several functions in the ArcPy mapping module that are specifically designed to manage layers within a data frame. These include the following:

- `AddLayer`—makes it possible to add a layer to a data frame within a map document using general placement options.

- `AddLayerToGroup`— makes it possible to add a layer to a group layer within a map document using general placement options.

- `InsertLayer`—makes it possible to add a layer to a data frame or to a group layer within a map document. It provides a more precise way of positioning the layer by using a reference layer.

- `MoveLayer`—makes it possible to move a layer to a specific location within a data frame or group layer within a map document.

- `RemoveLayer`—makes it possible to remove a layer from a map document.

- `UpdateLayer`—makes it possible to update the layer properties or just the symbology of a layer in a map document by extracting the information from a source layer.

These functions all must reference an already existing layer. It can be a layer file on disk, a layer from within the same map document, or a layer from a different map document. Thus, these functions do not perform the task of adding data to a map document, as Add Data does in ArcMap.

# 10.7 Fixing broken data sources

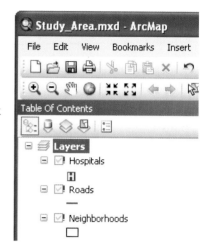

Consider the following scenario: You open an existing map document that you have not used in some time. Or perhaps a coworker has given you a disk or drive that has a project on it that contains map documents. When you open the map document, the layers in the ArcMap table of contents are shown with a red exclamation mark next to them and none of the layers are showing in the data frame. →

What happened? The link to the data source(s) has been lost. This can occur in a number of different scenarios:

- The map document was saved with full paths and the path to the data source has changed—for example, by moving it to a different drive.

- The map document was saved with relative paths, but the .mxd file has been moved relative to the data source.

- The names of data source files have been modified.

These broken data sources can be fixed within ArcMap as follows: right-click the layer, click Data > Repair Data Source, and browse to the correct data source. A few strategies can be used to prevent such broken data sources in the first place, including saving map documents with relative paths and proper file management in general.

Broken data sources are very common, and fixing them manually can be tedious. Scripting can be used to automate these corrections once the nature of the fix has been identified. Changes can be made to the map document without even opening it. Before examining these methods in more detail, a few definitions are in order:

- *Workspace*—a container for data. It can be a folder that contains shapefiles, a coverage, or a geodatabase—for example, mydata.

- *Workspace path*—the system path to a workspace. It includes the drive letter where the folder is located and any subfolders—for example, C:\mydata. For a file-based geodatabase, it includes the name of the geodatabase—for example, C:\mydata\project.gdb.

- *Dataset*—the feature class or table in the workspace. It is the actual name on disk, not the name displayed in the ArcMap table of contents. For a shapefile, it would be something like hospitals.shp. For a feature class in a geodatabase, it would be something like hospitals.

- *Data source*—the combination of workspace and dataset—for example, mydata\hospitals.shp or mydata\project.gdb\hospitals.

Prior to using a map document, you should check for broken data sources using the `ListBrokenDataSources` function. This `arcpy.mapping` function returns a Python list of layer objects within a map document or layer file that have broken connections to their original data source. The syntax is

```
ListBrokenDataSources(map_document_or_layer)
```

The following code illustrates how this function can be used to print the names of the layers in a map document that have broken data sources:

```
import arcpy
mapdoc = arcpy.mapping.MapDocument("CURRENT")
brokenlist = arcpy.mapping.ListBrokenDataSources(mapdoc)
for lyr in brokenlist:
    print lyr.name
del mapdoc
```

Running this code returns the names of the layers as they appear in the ArcMap table of contents. Instead of the name, the layer property `dataSource` can be used to see the current data source being referenced by the layer, as follows:

```
print lyr.dataSource
```

This lists the data sources that are broken and could help identify what the correct data sources should be. Keep in mind, however, that the `ListBrokenDataSources` function cannot identify what the correct data sources are—this can only be determined by a user with knowledge of the map documents and the data.

Once it is established that data sources need to be updated or fixed, the following methods can be applied to map documents, layers, or tables:

- `findAndReplaceWorkspacePaths` and `replaceWorkspaces`— use to perform a find-and-replace operation on the workspace path and the workspace, respectively. This assumes that the datasets are correct. For example, you can change C:\mydata\hospitals.shp to C:\newdata\hospitals.shp, but the name of the dataset (in this case, hospitals.shp) cannot be modified.

- `replaceDataSource`—use to perform a find-and-replace operation on the workspace and the dataset. You can modify both the workspace and the dataset, or just the dataset. For example, you can change C:\mydata\hospitals.shp to C:\mydata\newhospitals.shp.

The following methods work on three different classes: `MapDocument`, `Layer`, and `TableView` objects. In total, there are six different methods:

1. `MapDocument.findAndReplaceWorkspacePaths`

2. `MapDocument.replaceWorkspaces`

3. `Layer.findAndReplaceWorkspacePath`

4. `Layer.replaceDataSource`

5. `TableView.findAndReplaceWorkspacePath`

6. `TableView.replaceDataSource`

The syntax of `MapDocument.findAndReplaceWorkspacePaths` is

```
MapDocument.findAndReplaceWorkspacePaths(find_workspace_path, replace_ ➜
 ➜ workspace_path, {validate})
```

Running this code searches for and replaces the workspace paths of all layers and tables in a map document that share that workspace. For example, the following code replaces all the workspace paths in the current map document:

```
import arcpy
mapdoc = arcpy.mapping.MapDocument("CURRENT")
mapdoc.findAndReplaceWorkspacePaths("C:/mydata", "C:/newdata")
mapdoc.save()
del mapdoc
```

The methods in this group have an optional validation parameter. This parameter allows you to verify if a workspace or dataset is valid before changing its value. If `validate` is set to `True` (the default value) and the data source is valid, the data source will be updated. If the data source is not valid, it will remain pointing to the original data source. If `validate` is set to `False`, the workspace path or dataset does not have to be valid (that is, it does not already exist). This condition would be used when you want to update the data sources prior to the data being created.

When replacing workspace paths, you can replace all or part of a path. For example, if a workspace has simply moved from one drive to another, you can replace D:\ with C:\.

The `MapDocument.findAndReplaceWorkspacePaths` method works on multiple workspace types at once, including shapefiles, file geodatabases, and others. However, the workspace type cannot be modified. The `MapDocument.replaceWorkspaces` method can be used to modify both the workspace path and the workspace type—for example, from a folder containing shapefiles to a file geodatabase. The method works on only one workspace at a time but can be used multiple times if more than one workspace type needs to be replaced. The syntax of the `MapDocument.replaceWorkspaces` method is

```
MapDocument.replaceWorkspaces(old_workspace_path, old_workspace_type, ➤
➤ new_workspace_path, new_workspace_type, {validate})
```

For example, in the following code, references to shapefiles are redirected to feature classes in a file geodatabase:

```
import arcpy
mapdoc = arcpy.mapping.MapDocument("C:/mydata/project.mxd")
mapdoc.replaceWorkspaces("C:/mydata/shapes", "SHAPEFILE_WORKSPACE", ➤
➤ "C:/mydata/database.gdb", "FILEGDB_WORKSPACE")
mapdoc.save()
del mapdoc
```

Notice exactly what happened here. The workspace is changed, but not the dataset. For example, if the data source for a layer was C:\mydata\hospitals.shp, it has been modified to C:\mydata\database.gdb\hospitals. Because the type of workspace is specified, the .shp extension for the datasets is automatically removed. The example code assumes that feature classes with the exact same names as the shapefiles exist in the file geodatabase. Remember that paths are not case sensitive.

Valid workspace types are listed as follows:

- ACCESS_WORKSPACE
- ARCINFO_WORKSPACE
- CAD_WORKSPACE
- EXCEL_WORKSPACE
- FILEGDB_WORKSPACE
- OLEDB_WORKSPACE
- PCCOVERAGE_WORKSPACE
- RASTER_WORKSPACE
- SDE_WORKSPACE
- SHAPEFILE_WORKSPACE
- TEXT_WORKSPACE
- TIN_WORKSPACE
- VPF_WORKSPACE

Notice that "personal geodatabase" is not specifically included as a type—instead, ACCESS_WORKSPACE is used.

When workspaces in a map document are modified, there are a few things that may not work:

- Joins and relates associated with raster layers and stand-alone tables are not updated.

- Definition queries may no longer work because a slightly different SQL syntax is used—for example, between file geodatabases and personal geodatabases. A slight modification to the SQL statement would fix this problem.

- Label expressions may no longer work for the same reason.

The methods discussed so far work on map documents. However, data sources can also be modified for individual layers. The `Layer.findAndReplaceWorkspacePath` method works on a `Layer` object and performs a find-and-replace operation on the workspace path for a single layer in a map document or layer file. The syntax of this method is

```
Layer.findAndReplaceWorkspacePath(find_workspace_path, replace_workspace_
    path, {validate})
```

The following code modifies the workspace for a layer that references a particular feature class in a personal geodatabase. Only a portion of the full path of the data source is replaced—in this case, using a different personal geodatabase:

```
import arcpy
mapdoc = arcpy.mapping.MapDocument("C:/mydata/project.mxd")
lyrlist = arcpy.mapping.ListLayers(mapdoc):
for lyr in lyrlist:
    if lyr.supports("DATASOURCE"):
        if lyr.dataSource == "C:/mydata/database.gdb/hospitals":
            lyr.findAndReplaceWorkspacePath("database.gdb", "newdata.
    gdb")
mapdoc.save()
del mapdoc
```

The `Layer.findAndReplaceWorkspacePath` method assumes the dataset has not changed. The `replaceDataSource` method can be used to change both the workspace and the dataset. The syntax of this method is

```
Layer.replaceDataSource(workspace_path, workspace_type, dataset_name,
    {validate})
```

The following code replaces a specific data source. In this case, the value of the `dataSource` property is used to determine whether a layer should have its data source updated:

```
import arcpy
mapdoc = arcpy.mapping.MapDocument("C:/mydata/project.mxd")
lyrlist = arcpy.mapping.ListLayers(mapdoc):
for lyr in lyrlist:
    if lyr.supports("DATASOURCE"):
        if lyr.dataSource == "C:/mydata/hospitals.shp":
            lyr.replaceDataSource("C:/mydata/hospitals.shp", "SHAPEFILE_ ➔
➔ WORKSPACE", "C:/mydata/newhospitals.shp")
mapdoc.save()
del mapdoc
```

The `findAndReplaceWorkspacePath` and `replaceDataSource` methods also exist for `TableView` objects. The syntax for using these methods to work with single tables is very similar to the syntax for working with layers.

## 10.8  Working with page layout elements

Map scripting can also be used to work with page layout elements, including graphics, legends, pictures, text, and several others. Typical properties that can be changed include name, size, position, and sometimes other properties that vary with each element type.

Similar to map documents, layout elements cannot be created using scripting, so they have to already exist in a map document. The `ListLayoutElements` function can be used to identify which elements exist within the layout of a particular map document. The syntax of this function is

```
ListLayoutElements(map_document, {element_type}, {wild_card})
```

The `ListLayoutElements` function returns a Python list of elements. The optional parameter `element_type` can be used to limit the list of elements to only those of a specific type. The specific types of elements that can be used in scripting are as follows:

- DATAFRAME_ELEMENT
- GRAPHIC_ELEMENT
- LEGEND_ELEMENT
- MAPSURROUND_ELEMENT
- PICTURE_ELEMENT
- TEXT_ELEMENT

Each of these elements corresponds to a class in the `arcpy.mapping` module. Several of these elements are described in this section in a bit more detail. When getting started with layout elements, however, it can be useful to first create an inventory of what exists. For example, the following code creates a list of all layout elements and prints their name and type:

```
import arcpy
mapdoc = arcpy.mapping.MapDocument(r"C:\mydata\project.mxd")
elemlist = arcpy.mapping.ListLayoutElements(mapdoc)
for elem in elemlist:
    print elem.name & " " & elem.type
del mapdoc
```

The printout may look something like the following:

```
Legend LEGEND_ELEMENT
Alternating Scale Bar MAPSURROUND_ELEMENT
North Arrow MAPSURROUND_ELEMENT
Title TEXT_ELEMENT
Study Area DATAFRAME_ELEMENT
```

Notice that several different items are called `MAPSURROUND_ELEMENT`. Technically, any layout element that has an association with a data frame is a `MAPSURROUND_ELEMENT` object. This includes the north arrow, scale bar, and scale text. A legend element is also associated with a data frame, but since it has some unique properties, it is a separate element type.

Once it is determined what layout elements are available, a specific element can be selected by using: (1) the index number of the element, (2) the `element_type` parameter, or (3) the `wild_card` parameter. For example, to work with the title element, the following lines of code can be used to obtain a list with only the object that contains the title.

Using the index number directly:

```
title = arcpy.mapping.ListLayoutElements(mapdoc)[3]
```

Using the `element_type` parameter:

```
title = arcpy.mapping.ListLayoutElements(mapdoc, "TEXT_ELEMENT")[0]
```

Using the `wild_card` parameter:

```
title = arcpy.mapping.ListLayoutElements(mapdoc, "", "Title")[0]
```

In the case of the `element_type` and `wild_card` parameters, the `ListLayoutElements` function returns a list with only a single object. Using an index number of zero ([0]) on this list returns the object instead of a list.

> *Note: Not all elements have a default name, especially if they have been copied from other applications. To use these elements in scripting, the user has to first manually set the name in the map document.*

Once a specific element is referenced, various properties can be accessed, such as the element's name, type, height, and width, and the x,y coordinates of the element's anchor position. Other properties will vary with the type of element. For example, an important property of the `textElement` object is the `text` property.

For a text element, all properties are read/write, with the exception of the type. The following code modifies the text of the title in a page layout to a new string:

```
import arcpy
mapdoc = arcpy.mapping.MapDocument("C:/mydata/project.mxd")
title = arcpy.mapping.ListLayoutElements(mapdoc, "TEXT_ELEMENT")[0]
title.text = "New Study Area"
mapdoc.save()
del mapdoc
```

A few more examples of code follow to illustrate some unique properties that can be modified using scripting.

A `PictureElement` object has a `sourceImage` property, which represents the path to the image data source. The following code illustrates how this path can be modified:

```
import arcpy
mapdoc = arcpy.mapping.MapDocument("CURRENT")
elemlist = arcpy.mapping.ListLayoutElements(mapdoc, "PICTURE_ELEMENT")
for elem in elemlist:
    if elem.name == "photo1":
        elem.sourceImage = "C:/myphotos/newimage.jpg"
mapdoc.save()
del mapdoc
```

The `LegendElement` object has an `autoAdd` property, which controls whether a layer should be automatically added to the legend when a layer is added to a data frame using the `AddLayer` function. The following code illustrates how the `autoAdd` property can be modified to control which layer gets added:

```
import arcpy
mapdoc = arcpy.mapping.MapDocument("CURRENT")
df = arcpy.mapping.ListDataFrames(mapdoc)[0]
lyr1 = arcpy.mapping.Layer("C:/mydata/Streets.lyr")
lyr2 = arcpy.mapping.Layer("C:/mydata/Ortho.lyr")
legend = arcpy.mapping.ListLayoutElements(mxd, "LEGEND_ELEMENT")[0]
legend.autoAdd = True
arcpy.mapping.AddLayer(df, lyr1, "BOTTOM")
legend.autoAdd = False
arcpy.mapping.AddLayer(df, lyr2, "BOTTOM")
mapdoc.save()
del mapdoc
```

Another useful property of the `LegendElement` object is `items`, which returns a list of the names of the individual legend items. The `LegendElement` object also has one method, `adjustColumnCount`, which allows you to set the number of columns in the legend.

# 10.9 Exporting maps

The ArcPy mapping module has a number of exporting functions. They are similar to the ArcMap operation File > Export Map. There is a separate function for each format. The functions are as follows:

- `ExportToAI`
- `ExportToBMP`
- `ExportToEMF`
- `ExportToEPS`
- `ExportToGIF`
- `ExportToJPEG`
- `ExportToPDF`
- `ExportToPNG`
- `ExportToSVG`
- `ExportToTIFF`

These functions all work in a similar manner. The only required elements
for the export functions are a map document and the path and file name of
the output file. For example, the syntax of the ExportToJPEG function is
as follows:

```
ExportToJPEG (map_document, out_jpeg, {data_frame}, {df_export_width}, →
→ {df_export_height}, {resolution}, {world_file}, {color_mode}, {jpeg_→
→ quality}, {progressive})
```

The optional parameters represent the export options, which are also found
on the Export Map dialog box in ArcMap. For example, for the JPEG format,
these options look like the General and Format options shown in the two
figures.

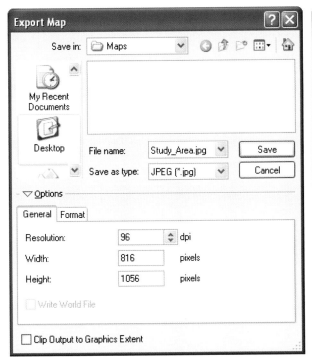
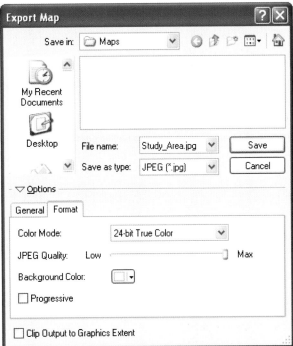

The dialog box options correspond directly to the parameters in the `ExportToJPEG` function. All these parameters have default values, and typically only selected parameters need to be set. The following code exports the page layout of a map document to a .jpg file, setting a resolution of 600 dpi:

```
import arcpy
mapdoc = arcpy.mapping.MapDocument("C:/project/study.mxd")
arcpy.mapping.ExportToJPEG(mapdoc, "C:/project/final.jpg", "", "", "", →
→ 600)
del mapdoc
```

Notice the use of empty strings ("") to skip several optional parameters.

One of the optional parameters in all export functions is the `data_frame` parameter. This parameter makes it possible to reference an individual `DataFrame` object to export, exporting just the data frame in Data View without any of the layout elements. By default, the page layout is used for export, including all data frames and layout elements.

Many of the other parameters will vary with the specific format selected.

## 10.10 Printing maps

In addition to exporting maps to files, the ArcPy mapping module contains a basic `PrintMap` function, which prints a specific data frame or map document to a printer or file. The syntax of this function is

```
PrintMap (map_document, {printer_name}, {data_frame}, {out_print_file})
```

The only required parameter is a map document. An optional `printer_name` parameter can be specified to represent the name of a printer on a local computer. If no printer is specified, the `PrintMap` function uses the printer that is saved with the map document or the default system printer if no printer is saved with the map document. An optional `data_frame` parameter can be used to reference a specific data frame—by default, the page layout is printed.

# 10.11 **Working with PDFs**

The PDF format has become widely used in the distribution of cartographic products. In addition to the `ExportToPDF` function, the ArcPy mapping module has a number of classes and functions to work with .pdf files. First, there is the `PDFDocument` class. This object allows for the manipulation of PDF documents, including the merging of pages, setting document behavior, and creating security settings. The syntax of the `PDFDocument` class is

```
PDFDocument(pdf_path)
```

The only parameter is a string that specifies the path and file name of the .pdf file. A `PDFDocument` object has only one property: `pageCount`, which is an integer for the number of pages. A `PDFDocument` object has five methods: `appendPages`, `insertPages`, `saveAndClose`, `updateDocProperties`, and `updateDocSecurity`.

There are two `PDFDocument` functions:

1. `PDFDocumentCreate`—creates an empty `PDFDocument` object in memory. The function receives a path and file name to determine the save location where a new PDF file will be created.

2. `PDFDocumentOpen`—returns a `PDFDocument` object from a PDF file on disk

These functions are often used to create a PDF map book. A number of separate .pdf files can be exported from map documents—for example, using the `DataDrivenPages` object discussed in the next section. These are appended in a newly created `PDFDocument` object and saved as a final PDF map book.

The following code creates an empty `PDFDocument` using the `PDFDocumentCreate` function, and appends three existing .pdf files into a single PDF. The `saveAndClose` method saves the resulting PDF, as follows:

```
import arcpy
pdfpath = "C:/project/MapBook.pdf"
pdfdoc = arcpy.mapping.PDFDocumentCreate(pdfpath)
pdfdoc.appendPages("C:/project/Cover.pdf")
pdfdoc.appendPages("C:/project/Map1.pdf")
pdfdoc.appendPages("C:/project/Map2.pdf")
pdfdoc.saveAndClose()
del pdfdoc
```

**>>> TIP**

The `PDFDocumentCreate` function does not actually create any blank PDF pages. In the example at left, the actual PDF pages come from existing PDF files. However, these pages could also be created in a script from a map document by using the `ExportToPDF` function.

# 10.12 Creating map books

ArcGIS has a set of tools to create a map book, which is simply a collection of pages printed together. A typical example of a map book includes an index page followed by individual maps. The index page shows the extent of the individual maps. An example map book is shown in the figure, including an index map and two of the many individual maps.

These map books can be created manually simply by printing each map separately. However, ArcMap contains a toolbar called Data Driven Pages to automate this process. More advanced map books require the use of scripting with the `DataDrivenPages` object in the ArcPy mapping module.

Automating the creation of map books using scripting requires that Data Driven Pages be enabled in the map document. This can be accomplished using the Data Driven Pages toolbar in ArcMap. On the Setup Data Driven Pages dialog box, a layer is selected that defines a series of extents—this layer is referred to as an index layer.

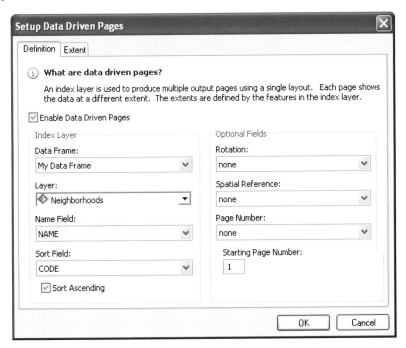

The `DataDrivenPages` object in the ArcPy scripting module can be used to access the properties and methods for managing the individual pages within the map document.

> **Note:** *A detailed explanation of how to work with Data Driven Pages and create map books is not provided in this book. For detailed explanations of these procedures, see "Creating a map book" and "Creating Data Driven Pages" on the Contents tab in ArcGIS Desktop Help (Mapping > Page Layouts).*

The exportToPDF method for the DataDrivenPages object can be used to create a map book in PDF format. This is not the same as the ExportToPDF function. The syntax of the exportToPDF method is as follows:

```
exportToPDF(out_pdf, {page_range_type}, {page_range_string}, {multiple_
  files}, {resolution}, {image_quality}, {colorspace}, {compress_vectors},
  {image_compression}, {picture_symbol}, {convert_markers}, {embed_fonts},
  {layers_attributes}, {georef_info})
```

The following code prints all the pages from a map document that has Data Driven Pages enabled to a PDF file and places an existing cover page in front:

```
import arcpy
pdfpath = "C:/project/MapBook.pdf"
pdfdoc = arcpy.mapping.PDFDocumentCreate(pdfpath)
mapdoc = arcpy.mapping.MapDocument("C:/project/Maps.mxd")
mapdoc.dataDrivenPages.exportToPDF("C:/project/Maps.pdf")
pdfdoc.appendPages("C:/project/Cover.pdf")
pdfdoc.appendPages("C:/project/Maps.pdf")
pdfdoc.saveAndClose()
del mapdoc
```

The Data Driven Pages toolbar and scripting can be used in combination to effectively produce map books. Some of the inherent behavior of Data Driven Pages such as page extents, scales, dynamic text, and the like are probably easiest to control on the Setup Data Driven Pages dialog box in ArcMap, although printing the pages and merging different PDF files is easiest to control using scripting.

## 10.13 Using sample mapping scripts

With the release of ArcGIS 10, Esri started making a number of script tools available to illustrate the use of ArcPy. These include a set of mapping script tools created as representative samples of how arcpy.mapping can be used to perform a variety of mapping tasks. These tools can be found in the Geoprocessing Model and Script Tool Gallery on the ArcGIS Resource Center.

> **Note:** To obtain the sample tools, go to http://resources.ArcGIS.com *and in the Search box, type **arcpy.mapping sample script tools**. This brings up a link to the sample tools.*

The sample tools consist of three different toolboxes: Cartography Tools (not to be confused with the existing Cartography system toolbox), Export and Printing Tools, and MXD and LYR Management Tools. Each toolbox contains a number of script tools, each of which references a Python script.

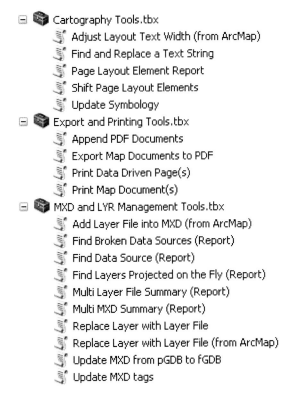

The sample tools include most of the functionality covered earlier in this chapter. Some of the scripts are quite short and simple. For example, the Print Map Document(s) tool prints the layout page of one or more map documents to a local printer. The tool dialog box, shown in the figure, allows you to select multiple map documents and select a local printer.

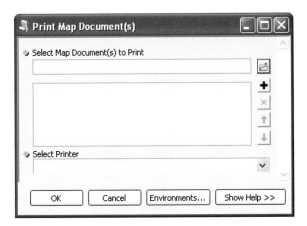

The Print Map Document(s) tool references the script file called
`PrintMXDs.py`, which is included in the files you get when downloading
the tools.

```
# Author:   ESRI
# Date:     July 5, 2010
# Version:  ArcGIS 10.0
# Purpose:  This script will print one or more map documents to a local printer.
#           The script is intended to run within a script tool.  There are two
#           parameters:
#                 1) Select Map Documents to Print,
#                 2) Select Output Printer (auto populated using a validation script)
#
#Notes: The print order of the MXDs is based on how they are entered.  The MXD
#       at the top of the list is first followed by those below it.

import arcpy, string
import arcpy.mapping as MAP

#Read input parameters from script tool
MXDList = string.split(arcpy.GetParameterAsText(0), ";")
printer = arcpy.GetParameterAsText(1)

#Loop through each MXD and print
for MXDPath in MXDList:
    MXD = MAP.MapDocument(MXDPath)
    MAP.PrintMap(MXD, printer)

    #Remove variable reference to file
    del MXD
```

The script tool uses the `GetParameterAsText` function to get the list of
map documents and the local printer from a user. You will learn about this
function in chapter 13 on creating custom tools. The script tool then uses
the `PrintMap` function in the `arcpy.mapping` module to print the map
documents. It can be useful to review sample scripts like this to get ideas
for your own scripts.

You can use these code examples as is, but you are encouraged to
modify them or use parts of the code in your own scripts.

>>> TIP

When modifying sample
scripts, you should first
save a copy of the script
since there is no Undo
button once you save your
changes to the script.

# Points to remember

- The `arcpy.mapping` module makes it possible to automate mapping tasks. A number of specific mapping classes and functions allow for the manipulation of map documents, data frames, layers, and page layouts.

- The functionality of the `arcpy.mapping` module reflects some of the typical workflows in ArcMap to produce cartographic output. Many procedures, however, are not part of the `arcpy.mapping` module because they lend themselves much more to the highly visual interface of ArcMap. For example, most of the layer symbology properties can be set only on the Layer Properties dialog box in ArcMap. The `arcpy.mapping` module can be used to automate certain repetitive tasks, such as updating the data sources for a large number of layers or replacing text in a large number of map documents.

- Map documents can be opened by referencing .mxd files on disk or by calling the map document currently in use. Map document properties can be accessed, modified, and saved. The `arcpy.mapping` module cannot create new map documents.

- Data frames within a map document can be accessed using the `ListDataFrames` function. Data frame properties can be accessed, modified, and saved.

- Layers within a data frame can be accessed using the `Layer` and `ListLayers` functions. Layer properties can be accessed, modified, and saved.

- Broken data sources in map documents can be identified using the `ListBrokenDataSources` function. Various methods exist to fix broken data sources for map document, layer, and table view objects. These methods can find and replace workspaces, workspace paths, and data sources.

- Individual elements on page layouts can be accessed and modified.

- Maps can be exported to various formats, including PDF, JPEG, and TIFF formats. Maps can also be printed to a local printer or to PDF format. When Data Driven Pages is enabled, scripting can be used to create a map book in PDF format.

# Chapter 11
## Debugging and error handling

## 11.1 Introduction

This chapter discusses debugging procedures and provides a review of the most common Python errors. Error-handling procedures are also discussed, including how to get the most out of `try-except` statements.

No matter how careful you are in writing code, errors are bound to happen. There are three main types of errors you will encounter in Python: *syntax errors*, *exceptions*, and *logic errors*. Syntax errors prevent code from running. With an exception, a script will stop running midprocess. A logic error means the script will run but produce undesired results.

## 11.2 Recognizing syntax errors

Syntax errors pertain to spelling, punctuation, and indentation. Common syntax errors result from misspelled keywords or variables, missing punctuation, and inconsistent indentation. See if you can spot the error in the following code:

```
import arcpy
from arcpy import env
env.workspace = "C:/Data/mydata.gdb"
fclist = arcpy.ListFeatureClasses()
for fc in fclist
    count = arcpy.GetCount_management(fc)
    print count
```

The colon (:) at the end of the first line of the `for` loop is missing. The following syntax error is displayed when this code is run in the Python window:

```
Parsing error SyntaxError: invalid syntax
```

PythonWin has a built-in checking process, which works somewhat like a spell checker in a word-processing application. The process is enabled by clicking the Check ⤴ button on the PythonWin Standard toolbar. This checks the current script file without running it.

Consider the preceding example code, which is shown in the figure.

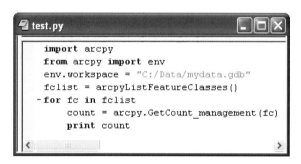

```
import arcpy
from arcpy import env
env.workspace = "C:/Data/mydata.gdb"
fclist = arcpyListFeatureClasses()
-for fc in fclist
    count = arcpy.GetCount_management(fc)
    print count
```

>>> TIP
Remember that you can make the line numbers visible in PythonWin by clicking View > Options > Editor on the menu bar, and then increasing the margin width for line numbers.

When you click the Check button, a message appears on the PythonWin status bar: `Failed to run script – syntax error – invalid syntax`. The cursor is also moved to the line where the first syntax error was detected—in this case, line 5, as shown in the figure.

Failed to run script - syntax error - invalid syntax                NUM      00005 017

Changing line 5 to the following code fixes the syntax error:

```
for fc in fclist:
```

Running this code produces the following message: `Python and the TabNanny successfully checked the file 'test.py'`. This means there are no syntax errors and the code will run when you click the Run button.

The Check button runs a syntax checker and TabNanny, which checks for inconsistent indentation and spacing. For example, the following block of code uses inconsistent indentation:

```
for fc in fclist:
    count = arcpy.GetCount_management(fc)
     print count
```

Using this code produces the following message:

```
Failed to check - syntax error - unexpected indent
```

Similar errors can occur if your indentation uses a combination of spaces and tabs. Especially when you copy from other sources, such as Microsoft Word, Microsoft PowerPoint, or PDF, blocks of code may appear to be indented correctly visually but may actually use a combination of spaces and tabs.

Consider the example code in the figure.

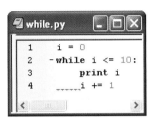

> **>>> TIP**
>
> Copying code from other sources such as Word and .pdf files will likely introduce other types of errors, including quotation marks. In general, therefore, it is not recommended to copy code from these types of files.

Visually, the block of code appears to be aligned, but TabNanny has underlined some of the whitespace in red. When a check is run, an error message appears, like the example in the figure.

Failed to check - syntax error - unindent does not match any outer indentation level　　　　NUM　　00004 011

To reveal the nature of the error, on the PythonWin menu bar, click View > Whitespace. The type of characters used for the indentation is displayed in the script window.

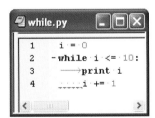

The arrow in the script window indicates the use of a tab, and the dots indicate spaces. Indentation should be consistent, so the tab should be replaced by four spaces.

> *Note: Although Python reports the line where a syntax error occurs, sometimes the actual syntax error occurs on a line that's above the one reported.*

> **>>> TIP**
>
> Using the TAB key in PythonWin results in four spaces by default. The spacing can be modified by clicking View > Options > Tabs and Whitespace from the menu bar. An actual tab is normally introduced only when copying from other applications.

# 11.3 Recognizing exceptions

Syntax errors are frustrating, but they are relatively easy to catch compared to other errors. Consider the following example that has the syntax corrected:

```
import arcpy
from arcpy import env
env.workspace = "C:/Data/mydata.gdb"
fclist = arcpy.ListFeatureClasses()
for fc in fclist:
    count = arcpy.GetCount_management(fc)
    print count
```

When you run the script again, it runs without a syntax error. But what if no count is printed? Is that an error? Perhaps the workspace is incorrect, or perhaps there are no feature classes in the workspace.

Rather than referring to these incidents as "errors," it is common for programming languages to discern between a normal course of events and something exceptional. There might be errors, but there might simply be events you might not expect to happen. These events are called *exceptions*. Exceptions refer to errors that are detected while the script is running. When an exception is detected, the script stops running unless the detection is handled properly. Exceptions are said to be *thrown*. If the exception is handled properly—that is, it is *caught*—the program can continue running. Examples of exceptions and proper error-handling techniques are covered later in this chapter.

# 11.4 Using debugging

When code results in exception errors or logic errors, you may need to look more closely at the values of variables in your script. This can be accomplished using a debugging procedure. Debugging is a methodological process for finding errors in your script. There are a number of possible debugging procedures, from very basic to more complex. Debugging procedures include the following:

- Carefully reviewing the content of error messages

- Adding print statements to your script

- Selectively commenting out code

- Using a Python debugger

Each of these approaches is reviewed in this section in more detail. Keep in mind that most of the time, debugging does not tell you *why* a script did not run properly, but it will tell you *where*—that is, on which line of code it failed. Typically, you still have to figure out why the error occurred.

## Carefully reviewing the content of error messages

Error messages generated by ArcPy are usually informative. Consider the following example:

```
import arcpy
arcpy.env.workspace = "C:/Data"
infcs = ["streams.shp", "floodzone.shp"]
outfc = "union.shp"
arcpy.Union_analysis(infcs, outfc)
```

This script carries out a union between two input feature classes, which are entered as a list. The result should be a new output feature class in the same workspace. The error message in PythonWin is as follows:

```
ExecuteError: Failed to execute. Parameters are not valid.
ERROR 000366: Invalid geometry type
Failed to execute (Union).
```

This is a specific error message produced by ArcPy, also referred to as an `ExecuteError` exception. The message is useful because it includes the statement: `Invalid geometry type`. Closer inspection of the input feature classes reveals that one of the inputs (streams.shp) is a polyline feature class, and the Union tool works with polygon features only. So the error message does not tell you exactly what is wrong (that is, it did not say that streams.shp is the geometry type polyline and that the Union tool does not accept this geometry type), but it points you in the right direction.

Not all error messages are as useful. Consider the following script:

```
import arcpy
arcpy.env.workspace = "C:/mydata"
infcs = ["streams.shp", "floodzone.shp"]
outfc = "union.shp"
arcpy.Union_analysis(infcs, outfc)
```

>>> **TIP**

When a specific error code is included in the error message, such as ERROR 000366, you can learn more about it in ArcGIS Desktop Help. In Help, go to Geoprocessing > Tool errors and warnings, and browse to the specific error by number.

This is, in fact, the same script, but it uses a different workspace (C:\ mydata), which does not exist. The error message in PythonWin is as follows:

```
Traceback (most recent call last):
  File "C:\Python27\ArcGIS10.1\Lib\site-packages\PythonWin\pywin\ ➤
➤ framework\scriptutils.py", line 325, in RunScript
    exec codeObject in __main__.__dict__
  File "C:\data\myunion.py", line 5, in <module>
    arcpy.Union_analysis(infcs, outfc)
  File "C:\Program Files (x86)\ArcGIS\Desktop10.1\arcpy\arcpy\ ➤
➤ analysis.py", line 574, in Union
    raise e
ExecuteError: Failed to execute. Parameters are not valid.
ERROR 000366: Invalid geometry type
Failed to execute (Union).
```

**Note:** *When the same code is run in the Python window in ArcGIS, a different error message results:*

```
Runtime error <class 'ArcGISscripting.ExecuteError'>: ERROR 000732: Input ➤
➤ Features: Dataset streams.shp #;floodzone.shp # does not exist or is not ➤
➤ supported
```

The PythonWin error message is rather misleading. It appears to suggest that the error is on line 5 of the code (where the union is carried out) and that there is an issue with the geometry. The error is, in fact, on line 2 where an invalid workspace is defined, and the invalid geometry message results from the fact that no feature classes could be obtained from the nonexistent workspace. Unfortunately, the error-reporting functions can't always report a more specific error message, such as `Workspace does not exist`.

Carefully examining error messages can be useful since they may, in fact, hold the answer to how to fix a problem. But don't stare yourself blind poring over them because the error may be something quite different and the error messages could prove misleading.

## Adding print statements to your script

When you have multiple lines of code that contain geoprocessing tools, it may not always be clear on which line an error occurred. In such cases, it may be useful to add `print` statements after each geoprocessing tool or other important steps to confirm they were run successfully. Consider the following code:

```
import arcpy
from arcpy import env
env.overwriteOutput = True
env.workspace = "C:/Data"
arcpy.Buffer_analysis("roads.shp", "buffer.shp", "1000 METERS")
print "Buffer completed"
arcpy.Erase_analysis("buffer.shp", "zone.shp", "erase.shp")
print "Erase completed"
arcpy.Clip_analysis("erase.shp", "wetlands.shp", "clip.shp")
print "Clip completed"
```

Even if the error message is cryptic and not informative, the `print` statements will illustrate which steps have been completed. The error can most likely be traced to the block of code just prior to the `print` statement that did not execute.

`Print` statements can be effective, but they are most useful when you already have a good idea of what might be causing the error. One of the downsides of using `print` statements is that they need to be cleaned up once the error has been fixed, which can be a substantial amount of work.

## Selectively commenting out code

You can selectively *comment out* code to see if removing certain lines eliminates the error. If your script has a typical sequential workflow, you would work from the bottom up. For example, the following code illustrates how the lower lines of code are commented out, using double number signs (##), to isolate the error:

```
import arcpy
from arcpy import env
env.overwriteOutput = True
env.workspace = "C:/Data"
arcpy.Buffer_analysis("roads.shp", "buffer.shp", "1000 METERS")
##arcpy.Erase_analysis("buffer.shp", "streams.shp", "erase.shp")
##arcpy.Clip_analysis("erase.shp", "wetlands.shp", "clip.shp")
```

As with adding print statements, this approach of commenting out lines of code does not identify why an error occurs, but only helps you to isolate where it occurs.

## Using a Python debugger

Another, more systematic, approach to debugging code is to use a Python debugger. A debugger is a tool that allows you to step through your code line by line, to place breakpoints in your code to examine the conditions at that point, and to follow how certain variables change throughout your code. Python has a built-in debugger module called pdb. It is a bit cumbersome because it lacks a user interface. However, Python editors such as IDLE and PythonWin include a solid debugging environment. In the next example that follows, the PythonWin debugger is used.

In PythonWin, you can turn the Debugger toolbar on and off by clicking View > Toolbars > Debugging from the menu bar.

The tools on the Debugger toolbar are briefly described in table 11.1.

**Table 11.1 Tools on the Debugger toolbar**

| | | |
|---|---|---|
| 𝟞𝟞 | Watch | Makes the Watch window visible, which allows you to keep track of the values of specifically defined variables in a script |
| | Stack view | Makes the Stack window visible, which keeps track of all variables in a script |
| ● | Breakpoint list | Makes the Breakpoint list window visible, which lists all the current breakpoints in a script |
| | Toggle Breakpoint | Turns a breakpoint on or off at the cursor location in the current script |
| | Clear All Breakpoints | Removes all breakpoints from the current script |
| | Step (or Step Into) | Runs the current line of code and moves to the next line, which can be in a different module, function, or method |
| | Step Over | Runs the current line of code, and if it includes a Python module, function, or method, it runs it, and then returns to the next line of code in the original script |
| | Step Out | Runs the current Python module, function, or method, and then returns to the next line of code in the original script |
| ▶ | Go | Runs a script until the next breakpoint or until the last line of code is reached |
| | Close | Stops the execution of code and exits the debugger to return to the script |

A typical debugging procedure in PythonWin is as follows:

1.  Check for any syntax errors and save the script.

2.  Run the script, but this time select a Debugging option on the Run Script dialog box—for example, "Step-through in the debugger".

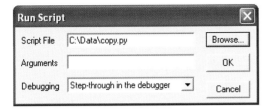

3.  Use the Step tool to go through your script line by line. Keep track of any error messages in the Interactive Window. The yellow arrow that appears in the window indicates the current line of code.

```
copy.py

1     import arcpy, os
2     from arcpy import env
3
4     #Allow for the overwriting of file geodatabases, if they already exist
5     env.overwriteOutput = True
6
7     #Set workspace to folder containing personal. geodatabase
8  ▷  env.workspace = "C:/Data"
9
10    #Identify personal geodatabases
11  ─for pgdb in arcpy.ListWorkspaces("", "Access"):
12        #Set workspace to current personal geodatabase
13        env.workspace = pgdb
14
15        #Create file geodatabase based on personal geodatabase
16        fgdb = pgdb[:4] + ".gdb"
17        arcpy.CreateFileGDB_management(os.path.dirname(fgdb), os.path.basename(fgdb))
```

4.  Use the Step Over and Step Out tools to skip ahead. For example, if the current line of code contains a call to a different module, function, or method, using the Step tool results in stepping into that procedure. Using the Step Over tool will run the line of code without stepping into that procedure—you are stepping over the details of that procedure. Once you step into a procedure, you can use the Step Out tool to step out of the procedure without stepping through the rest of the lines of code. This allows you to fast-forward and return to the next line of code that called the procedure.

5. While stepping through the script code, open the Watch and Stack windows to keep track of variables. The Stack window keeps track of all the variables and the Watch window keeps track of only the variables you specify manually.

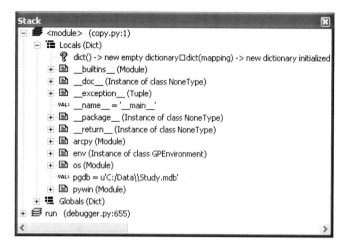

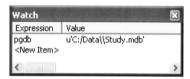

6. When you find an error, use the Close tool to stop the execution of the script. Then fix the error and run the script again.

When scripts get longer, going through a script line by line can be a bit cumbersome. Instead, you can place breakpoints in the script using the Toggle Breakpoint tool. The next time you run the script, you can select the "Run in the debugger" option.

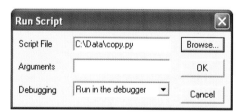

Then the debugger will stop the script at only the predefined breakpoints and run the lines in between breakpoints in one step instead of stopping at every line.

```
🗎 copy.py                                                                          [_][□][✕]
 1      import arcpy, os
 2      from arcpy import env
 3
 4      #Allow for the overwriting of file geodatabases, if they already exist
 5      env.overwriteOutput = True
 6
 7      #Set workspace to folder containing personal. geodatabase
 8 ◈    env.workspace = "C:/Data"
 9
10      #Identify personal geodatabases
11    ┌ for pgdb in arcpy.ListWorkspaces("", "Access"):
12          #Set workspace to current personal geodatabase
13          env.workspace = pgdb
14
15          #Create file geodatabase based on personal geodatabase
16 ⬤        fgdb = pgdb[:4] + ".gdb"
17          arcpy.CreateFileGDB_management(os.path.dirname(fgdb), os.path.basename(fgdb))
```

The breakpoints can be turned on and off by placing the cursor in the desired line of code and using the Toggle Breakpoint tool. To clear all the breakpoints in the current script, use the Clear All Breakpoints tool.

# 11.5 Using debugging tips and tricks

Following are some general tips and tricks that will help you to debug your scripts:

- Remember that ArcGIS for Desktop applications often place a lock on a file, which may prevent a script from overwriting the file.

- When working with very large files, first try your code on a small file with similar properties.

- Watch where the values of variables are changing by inserting print statements or breakpoints in the code.

- Place breakpoints inside blocks of code where repetition should be occurring.

- If PythonWin does not stop running while you are debugging, you can interrupt the code by right-clicking the PythonWin icon in the notification area, at the far-right corner of the taskbar, and then clicking "Break into running code". This will result in a KeyboardInterrupt exception in the Interactive Window without closing PythonWin. ➜

# 11.6 Error handling for exceptions

Although debugging procedures can contribute to writing correct code, exception errors are still likely to occur in your scripts. Exceptions refer to errors that are detected as the script is running. One key reason for this is that many scripts rely on user input, and you can't always control the input other users will provide. Well-written scripts, therefore, include error-handling procedures to handle exceptions. Error-handling procedures are written to avoid having a script fail and not provide meaningful feedback.

To handle exceptions, you could use conditional statements to check for certain scenarios, which is analogous to using an `if` statement. You have already encountered some in previous chapters. For example, the existence of a path can be determined in Python using a built-in Python function such as `os.path.exists`. For catalog paths, you can use the `Exists` function to determine whether data exists. For example, the following code determines whether a shapefile exists:

```
import arcpy
from arcpy import env
env.workspace = "C:/Data"
shape_exist = arcpy.Exists("streams.shp")
print shape_exist
```

The `Exists` function can be used for feature classes, tables, datasets, shapefiles, workspaces, layers, and files in the current workspace. The function returns a Boolean value indicating whether the element exists.

Besides determining whether data exists, you can determine whether the data is the right type by using the `Describe` function. For example, if your script requires a feature class, you can use the `datsetType` property to determine whether it is a feature class.

Writing conditional statements for every possible error is tedious. And it is impossible to foresee every error. In the example code earlier in this section, you would have to check the following: (1) whether the workspace is valid, (2) whether there is at least one feature class in the workspace, and (3) whether there is a feature class with at least one feature. This could easily double the code in the script.

There are two strategies to check for errors and report them in a meaningful manner:

1.  Use Python exception objects inside `try-except` statements.

2.  Report messages using the ArcPy messaging functions.

A powerful alternative to conditional statements is Python exception objects. When Python encounters an error, it *raises*, or *throws*, an exception. This typically means the script stops running. If such an exception object is not handled, or *caught*, the script terminates with a runtime error, sometimes also referred to as a *traceback*.

Consider the simple example of trying to divide by zero. In the Python window, it results in a runtime error, as shown in the figure.

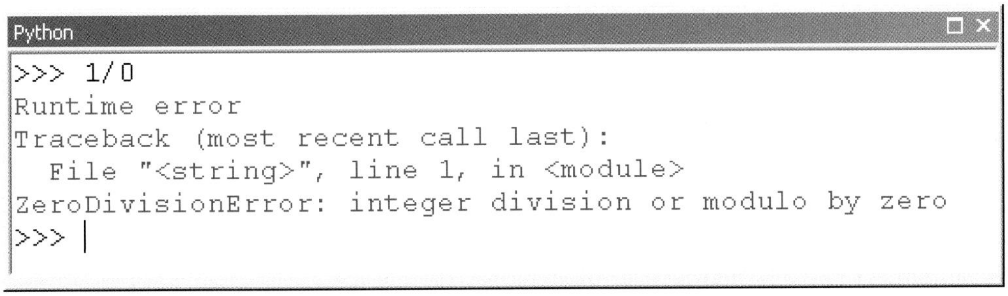
Python
>>> 1/0
Runtime error
Traceback (most recent call last):
File "<string>", line 1, in <module>
ZeroDivisionError: integer division or modulo by zero
>>>

The following sections will illustrate how exceptions are raised and how the `try-except` statement can be used to effectively trap errors.

## 11.7 Raising exceptions

Exceptions are raised automatically when something goes wrong. You can also raise exceptions yourself by using the `raise` statement. You can raise a generic exception using the `raise Exception` statement, as follows:

```
>>> raise Exception
Runtime error
Exception
```

You can also add a specific message, as follows:

```
>>> raise Exception("invalid workspace")
Runtime error
Exception: invalid workspace
```

There are many different types of exceptions. You can view all of them by importing the `exceptions` module and using the `dir` function to list all of them:

```
>>> import exceptions
>>> dir(exceptions)
```

Running this code results in a long printout (not shown in entirety here):

```
['ArithmeticError', 'AssertionError', 'AttributeError', 'BaseException', ➜
➜ 'BufferError', 'BytesWarning' ...]
```

Each of these exceptions can be used in the `raise` statement. For example:

```
>>> raise ValueError
Runtime error
ValueError
```

This example is a *named* exception—that is, the specific exception is called by name. Named exceptions allow a script to handle specific exceptions in different ways, which can be beneficial. Using the generic `Exception` is referred to as an *unnamed* exception.

A complete description of built-in Python exceptions can be found in the Python documentation. In the documentation, go to Library Reference and browse to section 6, "Built-in Exceptions." It also includes a hierarchy of errors: for example, `ZeroDivisionError` is one of several types of arithmetic errors (`ArithmeticError`).

> **>>> TIP**
>
> The Python documentation is installed in the same program folder as ArcGIS. For a typical installation, you can find the documentation by clicking the Start button on the taskbar, and then, on the Start menu, clicking All Programs > ArcGIS > Python 2.7 > Python Manuals.

---

Python v2.7 documentation » The Python Standard Library »                 previous | next | modules | index

# 6. Built-in Exceptions

Exceptions should be class objects. The exceptions are defined in the module **exceptions**. This module never needs to be imported explicitly: the exceptions are provided in the built-in namespace as well as the **exceptions** module.

For class exceptions, in a **try** statement with an **except** clause that mentions a particular class, that clause also handles any exception classes derived from that class (but not exception classes from which *it* is derived). Two exception classes that are not related via subclassing are never equivalent, even if they have the same name.

The built-in exceptions listed below can be generated by the interpreter or built-in functions. Except where mentioned, they have an "associated value" indicating the detailed cause of the error. This may be a string or a tuple containing several items of information (e.g., an error code and a string explaining the code). The associated value is the second argument to the **raise** statement. If the exception class is derived from the standard root class **BaseException**, the associated value is present as the exception instance's **args** attribute.

User code can raise built-in exceptions. This can be used to test an exception handler or to report an error condition "just like" the situation in which the interpreter raises the same exception; but beware that there is nothing to prevent user code from raising an inappropriate error.

The built-in exception classes can be sub-classed to define new exceptions; programmers are encouraged to at least derive new exceptions from the **Exception** class and not **BaseException**. More information on defining exceptions is available in the Python Tutorial under *User-defined Exceptions*.

The following exceptions are only used as base classes for other exceptions.

*exception* **BaseException**

> The base class for all built-in exceptions. It is not meant to be directly inherited by user-defined classes (for that use **Exception**). If **str()** or **unicode()** is called on an instance of this class, the representation of the argument(s) to the instance are returned or the empty string when there were no arguments. All arguments are stored in **args** as a tuple.

# 11.8 **Handling exceptions**

Exceptions in a script can be handled using a `try-except` statement. Handling exceptions is often called *trapping*, or *catching*, the exceptions. When an exception is properly handled, the script does not produce a runtime error but instead reports a more meaningful error message to the user. This means the error is trapped, or caught, before it can cause a runtime error.

Consider the following script that divides two user-supplied numbers:

```
x = input("First number: ")
y = input("Second number: ")
print x/y
```

The script will work fine until zero (0) is entered as the second number, resulting in the following error message:

```
First number: 100
Second number: 0
Traceback (most recent call last):
  File " division.py", line 3, in <module>
    print x/y
ZeroDivisionError: integer division or modulo by zero
```

The `try-except` statement can be used to trap this exception and provide additional error handling, as follows:

```
try:
    x = input("First number: ")
    y = input("Second number: ")
    print x/y
except ZeroDivisionError:
    print "The second number cannot be zero."
```

Notice the structure of the `try-except` statement. The first line of code consists of only the `try` statement, followed by a colon (:). Next is a block of indented code with the procedure you want to carry out. Then comes the `except` statement, which includes a specific exception, followed by a colon (:). Next is a block of indented code that will be carried out if the specific exception is raised. The exception `ZeroDivisionError` is a named exception.

In this example, a simple `if` statement might have been more effective to determine whether the value of y is zero (0). However, for more elaborate code, you might need many such `if` statements, and a single `try-except` statement will be sufficient to trap the error.

Multiple `except` statements can be used to catch different named exceptions. For example:

```
try:
    x = input("First number: ")
    y = input("Second number: ")
    print x/y
except ZeroDivisionError:
    print "The second number cannot be zero."
except TypeError:
    print "Only numbers are valid entries."
```

You can also catch multiple exceptions with a single block of code by specifying them as a tuple:

```
except (ZeroDivisionError, TypeError):
    print "Your entries were not valid."
```

In this case, the error handling is not specific to the type of exception and only a single message is printed, no matter what type of error caused the exception. The error message, in this case, is less specific because it describes several exceptions.

The exception object itself can also be called by providing an additional argument:

```
except (ZeroDivisionError, TypeError) as e:
    print e
```

Running this code allows you to catch the exception object itself and you can print it to see what happened rather than printing a custom error message.

It can be difficult sometimes to predict all the types of exceptions that might occur. Especially in a script that relies on user input, you may not be able to foresee all the possible scenarios. So to catch all the exceptions, no matter what type, you can simply omit the exception class from the `except` statement, as follows:

```
try:
    x = input("First number: ")
    y = input("Second number: ")
    print x/y
except Exception as e:
    print e
```

In this example, the exception is unnamed.

The `try-except` statement can also include an `else` statement, similar to a conditional statement. For example:

```
while True:
    try:
        x = input("First number: ")
        y = input("Second number: ")
        print x/y
    except:
        print "Please try again."
    else:
        break
```

In this example, the `try` block of code is repeated in a `while` loop when an exception is raised. The loop is broken by the `break` statement in the `else` statement only when no exception is raised.

One more addition to the `try-except` statement is the `finally` statement. Whatever the result of previous `try`, `except`, or `else` blocks of code, the `finally` block of code will always be executed. This block typically consists of clean-up tasks and could include checking in licenses or deleting references to map documents.

# 11.9 **Handling geoprocessing exceptions**

So far, the exceptions raised have been quite general. A Python script can, of course, fail for many reasons that are not specifically related to a geoprocessing tool as the previous examples illustrate. However, because errors related to geoprocessing tools are somewhat unusual in nature, they warrant more attention.

You can think of errors as falling into two categories: geoprocessing errors and everything else. When a geoprocessing tool writes an error message, ArcPy generates a system error. Specifically, when a geoprocessing tool fails to run, it throws an `ExecuteError` exception, which can be used to handle specific geoprocessing errors. It is not one of the built-in Python exception classes, but it is generated by ArcPy and thus the `arcpy.ExecuteError` class has to be used.

Consider this example:

```
import arcpy
arcpy.env.workspace = "C:/Data"
in_features = "streams.shp"
out_features = "streams.shp"
try:
    arcpy.CopyFeatures_management(in_features, out_features)
except arcpy.ExecuteError:
    print arcpy.GetMessages(2)
except:
    print "There has been a nontool error."
```

The Copy Features tool generates an error because the input and output feature classes cannot be the same, as follows:

```
Failed to execute. Parameters are not valid.
ERROR 000725: Output Feature Class: Dataset C:/Data\zip.shp already  ➡
➡ exists.
Failed to execute (CopyFeatures).
```

In the example code, the first `except` statement traps any geoprocessing errors, and the second `except` statement traps any nongeoprocessing errors. This example illustrates how both named and unnamed exceptions can be used in the same script. It is important to first check the named exceptions, such as `except arcpy.ExecuteError`, and then the unnamed exceptions. If the unnamed exceptions were checked first, the statement would catch all exceptions, including any `arcpy.ExecuteError` exceptions. This would mean you would never know whether a named exception (that you put in the script) occurred or not.

In larger scripts, it can be difficult to determine the precise location of an error. You can use the Python `traceback` module to isolate the location and cause of an error.

The `traceback` structure is as follows:

```
try:
    import arcpy
    import sys
    import traceback
    <block of code including geoprocessing tools>
except:
    tb = sys.exc_info()[2]
    tbinfo = traceback.format_tb(tb)[0]
    pymsg = "PYTHON ERRORS:\nTraceback info:\n" + tbinfo + "\nError ➤
➤ Info:\n" + str(sys.exc_type) + ":" + str(sys.exc_value) + "\n"
    arcpy.AddError(pymsg)
    msgs = "ArcPy ERRORS:\n" + arcpy.GetMessages(2) + "\n"
    arcpy.AddError(msgs)
    print pymsg + "\n"
    print msgs
```

In this code, two types of errors are trapped: geoprocessing errors and all other types of errors. The geoprocessing errors are obtained using the ArcPy `GetMessages` function. The errors are returned for use in a script tool (`AddError`) and also printed to the standard Python output (`print`). All other types of errors are retrieved using the `traceback` module. Some formatting is applied and the errors are returned for use in a script tool and printed to the standard Python output.

Following is one more example of a `try-except` statement, using the `finally` statement. In this example, a custom exception class is created to handle a license error. A license is checked out in the `try` code block and the license is checked in as part of the `finally` code block. This ensures the license is checked in, no matter the outcome of running the earlier code blocks, as follows:

```
class LicenseError(Exception):
    pass
import arcpy
from arcpy import env
try:
    if arcpy.CheckExtension("3D") == "Available":
        arcpy.CheckOutExtension("3D")
    else:
        raise LicenseError
    env.workspace = "C:/raster"
    arcpy.Slope_3d("elevation", "slope")
except LicenseError:
    print "3D license is unavailable"
except:
    print arcpy.GetMessages(2)
finally:
    arcpy.CheckInExtension("Spatial")
```

Using the `try-except` statement for error trapping is very common. The `ExecuteError` exception class is useful, but in practice, most scripts rely on the simple but effective `try-except` statement without using specific exception classes.

Sometimes, you will see an entire script wrapped in a `try-except` statement. It would look something like the following structure in which the `try` code block could contain hundreds of lines of code:

```
try:
    import arcpy
    import traceback
    ##multiple lines of code here

except:
    tb = sys.exc_info()[2]
    tbinfo = traceback.format_tb(tb)[0]
    pymsg = "PYTHON ERRORS:\nTraceback info:\n" + tbinfo + "\nError ➔
➔ Info:\n" + str(sys.exc_info()[1])
    msgs = "ArcPy ERRORS:\n" + arcpy.GetMessages(2) + "\n"
    arcpy.AddError(pymsg)
    arcpy.AddError(msgs)
```

# 11.10 Using other error-handling methods

In addition to a `try-except` statement for trapping errors in scripts, several other error-handling methods can be used. Some of them are covered in earlier chapters but warrant further mention here:

- Validating table and field names using the `ValidateTableName` and `ValidateFieldName` functions, respectively (chapter 7).

- Checking for licenses for products using the `CheckProduct` function and for extensions using the `CheckExtension` function (chapter 5).

- Checking for schema locks—many geoprocessing tools will not run properly if schema locks exist on the input.

# 11.11 Watching for common errors

Following are a number of common errors to look out for when you are scanning your scripts and examining your data.

## Common Python code errors

- Simple spelling mistakes

- Forgetting to import modules, such as `arcpy`, `os`, or `sys`

- Case sensitivity—for example, `mylist` versus `myList`

- Paths—for example, using a single backslash (\), such as C:\Data\streams.shp

- Forgetting colons (:) after statements (`for`, `while`, `else`, `try`, `except`)

- Incorrect or inconsistent indentation

- Conditional ( = = ) versus assignment ( = ) statements

## Common geoprocessing-related errors

- Forgetting to determine whether data exists. A small typo in the name of a workspace or feature class will cause a tool to fail. Always double-check that the inputs to a script exist.

- Forgetting to check for overwriting output. The default setting is not to overwrite outputs, so unless this option is specifically cleared, a tool that attempts to overwrite output will not run. A very common scenario is to run a script and it works, but when you run it a second time, it fails—fixing this could be as simple as setting the `overwriteOutput` property of the `env` class to `True`.

- Data is being used in another application. You may be trying to run a script, but it will not run because you are also using the data in ArcMap or ArcCatalog—this is very common because often you are exploring the data that is going to be used in the script. Closing these applications and trying the script again may resolve a script error.

- Not checking the properties of parameters and objects returned by tools. For example, it may sound logical that the Get Count tool produces a count—that is, a number. It actually returns a result object that is printed to the Results window, so you have to use the `getOutput` method to obtain this count. Similarly, distinctions between feature classes and feature layers may seem somewhat trivial, but they may be just the difference between proper tool execution and failure. Carefully examine tool syntax and determine the exact nature of the inputs and outputs.

It is worth noting that many geoprocessing-related errors can be prevented when using script tools. Building a script tool includes validation for preventing invalid parameters. This is covered in chapter 13.

Some of these suggestions may appear rather rudimentary, but the solutions can often be simple if you only knew where to look for them. The syntax of a good Python geoprocessing script is often relatively simple, which is part of the beauty of using Python.

**>>> TIP**

Remember that geoprocessing scripts don't have to follow Python coding logic alone. They must also obey the rules of the ArcGIS geoprocessing framework.

# Points to remember

- Errors in geoprocessing scripts are bound to happen. Although syntax errors are relatively easy to catch, your script may contain other types of errors that prevent proper script execution. Scripts can be made more robust by incorporating error-handling procedures.

- Various debugging methods exist. Relatively simple approaches include carefully examining error messages, adding print statements to the code to review intermediate results, and selectively commenting out code. If these methods are not sufficient to identify and fix errors, a Python debugger can be used such as the PythonWin Debugger. A debugger allows you to carefully step through the code line by line to review error messages and examine the state of variables. Breakpoints can be added to step through larger blocks of code.

- Any debugging procedure will typically identify where the error occurs but not exactly why it occurs. It is therefore good practice to always be aware of common errors, including Python coding errors and ArcGIS geoprocessing errors.

- Basic error-handling procedures include checking whether data exists, determining whether data inputs are the right type, checking for licenses and extensions, and validating table and field names. Typically, an `if` statement is used for this type of error handling.

- It is nearly impossible to anticipate every possible type of errors, and code that checks for such errors would become too cumbersome to write. Whenever something goes wrong in a script, an exception is automatically raised. These exceptions can be trapped using a `try-except` statement. This type of statement makes it possible to identify the type of error or else specific errors. Customized error-handling procedures can be implemented based on the nature of the error. Additional statements, including `else` and `finally`, can be added to the `try-except` statement to ensure efficient error trapping.

- Error messages can be very useful for identifying the nature of the error and how to fix the script. These include both general Python messages and error messages resulting from the ArcPy `ExecuteError` exception class.

# Chapter 12
## Creating Python functions and classes

## 12.1 Introduction

This chapter describes how to create custom functions in Python that can be called from elsewhere in the script or from another script. Custom functions make it easy to reduce the code you have written to carry out procedures. Functions are organized into modules, and modules can be organized into a package. ArcPy itself is a collection of custom modules and functions organized into a package. By creating custom functions, you can organize your code into logical parts and reuse frequently needed procedures. This chapter also describes how to create custom classes in Python, which makes it easier to group together functions and variables.

## 12.2 Creating functions

Functions are small blocks of code that perform a specific task. Python itself has a great number of built-in functions and the ArcPy site package contains a large number of functions, including all the geoprocessing tools in ArcGIS. You will use many built-in functions in a typical Python script, and you can import additional functionality from other modules, including ArcPy. Consider the `random` module, for example. You can import this module for access to a number of different functions. The following code generates a random integer between 1 and 100:

```
import random
x = random.randint(1,100)
print x
```

The code to generate a random number has already been written, and this code can now be freely used by anyone who needs it. The code of the random module can be found in a file called random.py and is located in the Python Lib folder. In a typical installation of Python 2.7 as part of the ArcGIS 10.1 installation, the path is: C:\Python27\ArcGIS10.1\Lib\random. py. You can open this script in a Python editor like PythonWin and examine the code. Inside the code, you will find a reference to the randint function, as shown in the figure.

```
random.py                                                            _ □ X

223
224  -      def randint(self, a, b):
225              """Return random integer in range [a, b], including both end points.
226              """
227
228              return self.randrange(a, b+1)
229
```

In this example, the randint function calls another function called randrange. The random module contains a number of different functions and some of them are closely related. The point here is that the code to generate random numbers has already been written and shared with the Python user community. So whenever your script needs a random number, you don't have to write the code yourself. You can import the random module and use any of its functions.

In addition to using existing functions, you can create your own functions that can be called from within the same script or from other scripts. Once you write your own functions, you can reuse them whenever needed. This makes code more efficient since there is no need to write the same task over and over.

Python functions are defined using the def statement, as shown in the figure. The def statement contains the name of the function, followed by any arguments in parens. The syntax of the def statement is

```
def <functionname>(<arguments>):
```

There is a colon (:) at the end of the statement, and the code following a def statement is indented the same as any block of code. This indented block of code is the function definition.

For example, consider the script helloworld.py as follows:

```
def printmessage():
    print "Hello world"
```

In this example, the function `printmessage` has no parameters, but most functions use parameters to pass values. Elsewhere in the same script, you can call this function directly, as follows:

```
printmessage()
```

Typically, functions are quite a bit more elaborate. Consider the following example: You want to create a list of the names of all the fields in a table or a feature class. There is no function in ArcPy that does this. However, the `ListFields` function allows you to create a list of the fields in a table, and you can then use a `for` loop to iterate over the items in the list to get the names of the fields. The list of names can be stored in a list object. The code is as follows:

```
import arcpy
arcpy.env.workspace = "C:/Data"
fields = arcpy.ListFields("streams.shp")
namelist = []
for field in fields:
    namelist.append(field.name)
```

Now, say you anticipate that you will be using these lines of code quite often—in the same script or in other scripts. You can simply copy the lines of code, paste them where they are needed, and make any necessary changes. For example, it is likely you will need to replace the parameter `"streams.shp"` with the feature class or table of interest.

Instead of copying and pasting code, you can define a custom function to carry out the same steps. First, you need to give the function a name—for example, `listfieldnames`. The following code defines the function:

```
def listfieldnames():
```

You can now call the function from elsewhere in the script by name. In this example, when calling the function, you want to pass a value to the function—that is, the name of a table or a feature class. To make this possible, the function needs to include a parameter to receive these values. The parameter needs to be included in the definition of the function, as follows:

```
def listfieldnames(table):
```

Following the `def` statement is an indented block of code that contains what the function actually does. This is identical to the previous lines of code, but now the hard-coded value of the feature class is replaced by the parameter of the function:

```
def listfieldnames(table):
    fields = arcpy.ListFields(table)
    namelist = []
    for field in fields:
        namelist.append(field.name)
```

The last thing needed is a way for the function to pass values, also referred to as *returning* values. This is necessary to ensure that the function not only creates the list of names, but also returns the list so it can be used by any code that calls the function. This is accomplished using a return statement. The completed description of the function is as follows:

```
def listfieldnames(table):
    fields = arcpy.ListFields(table)
    namelist = []
    for field in fields:
        namelist.append(field.name)
    return namelist
```

Once a function is defined, it can be called directly from within the same script, as follows:

```
fieldnames = listfieldnames("C:/Data/hospitals.shp")
```

Running the code returns a list of the fields in a table using the function previously defined. Notice that the new function `listfieldnames` can be called directly, since it is defined in the same script.

The example function used a parameter called `table`, which makes it possible to pass a value to the function. The parameter is also referred to as an argument. A function can use more than one parameter, and parameters can be made optional. The arguments for optional parameters should be ordered so that the required ones are listed first, followed by the optional ones. Arguments can be made optional by specifying default values.

Creating functions can be beneficial in a number of ways:

- If a task is to be used many times, creating a function can reduce the amount of code you need to write and manage. The actual code that carries out the task is written only once as a function, and from that point on, you can call this custom function as needed.

- Creating functions can reduce the clutter caused by multiple itera-tions. For example, if you wanted to create lists of the field names for all the feature classes in all the geodatabases in a list of workspaces, it would quickly create a relatively complicated set of nested `for` loops. Creating a function for creating a list of field names removes one of these `for` loops and places it in a separate function.

- Complex tasks can be broken into smaller steps. By defining each step as a function, the complex task does not appear so complex anymore. Well-defined functions are a good way to organize longer scripts.

Custom functions can be called not only directly from the same script, but also from other scripts, which is covered in the next section.

# 12.3 Calling functions from other scripts

Once functions are created in a script, they can be called from another script by importing the script that contains the function. For relatively complex functions, it is worthwhile to consider making them into separate scripts or script tools, especially if they are needed on a regular basis. So rather than defining a function within a script, the function becomes a script in itself that can be called from other scripts.

Consider the earlier example of the helloworld.py script:

```
def printmessage():
    print "Hello world"
```

The `printmessage` function can be called from another script by importing the helloworld.py script. For example, the script print.py does it, as follows:

```
import sys
import os
import helloworld
helloworld.printmessage()
```

The script print.py imports the `helloworld` module. A module name is equal to the name of the script minus the .py extension. The function is called using the regular syntax to call a function—that is, `<module>.<function>`.

In the example script, the `helloworld` module is imported into the print.py script. Notice that there is no path associated with the module, but just the name itself, which is the name of another script. So, the `import` statement causes Python to look for a file named helloworld.py. No paths can be used in the `import` statement, and thus it is important to recognize where Python looks for modules.

The first place Python looks for modules is the current folder, which is the folder where the print.py script is located. The current folder can be obtained using the following code, where `sys.path` is a list of system paths:

```
import sys
print sys.path[0]
```

The current folder can also be obtained using the `os` module, as follows:

```
import os
print os.getcwd()
```

Next, Python looks at all the other system paths that have been set during the installation or subsequent configuration of Python itself. These paths are contained in an environment settings variable called PYTHONPATH. This variable can be set to a list of paths that will be added to the beginning of the `sys.path` list. To view a complete list of these paths, use the following code:

```
import sys
print sys.path
```

In a typical scenario, the list will include paths to both the Python installation and the ArcGIS installation. The list will include paths like the following:

```
C:\Python27\ArcGIS10.1
C:\Python27\ArcGIS10.1\Lib
C:\Python27\ArcGIS10.1\Lib\site-packages
C:\Program Files\ArcGIS\Desktop10.1\bin
C:\Program Files\ArcGIS\Desktop10.1\arcpy
C:\Program Files\ArcGIS\Desktop10.1\ArcToolbox\Scripts
```

**Note:** *The list of paths will vary based on how ArcGIS and Python are installed and the versions of the software.*

What if the module you want to import is in a different folder—that is, not in the current folder of the script or in any of the folders in `sys.path`? You have two options, as follows:

1.  Use a path configuration file (.pth).

    The most convenient way to access a module in a different folder is to add a path configuration file to a folder that is already part of `sys.path`. It is common to use the site-packages folder—for example, C:\Python27\ArcGIS10.1\lib\site-packages. A path configuration file has a .pth extension and contains the path(s) that will be appended to `sys.path`. This file can be created using a basic text editor, and each line must contain a single path. When ArcPy is installed as part of the ArcGIS installation, a path configuration file called `Desktop10.1.pth` is placed in the site-packages folder of Python. The file itself looks like the example in the figure.

The path configuration file makes all the modules located in the specific folders available to Python.

You can create a .pth file yourself if you commonly work with scripts that are located in different folders. For example, if the modules you want to import are in the folder C:\Sharedscripts, you would create a .pth file and place it in the Python site-packages folder. The file itself would look like the example in the figure.

**Note:** *The paths in the path configuration file use a backslash (\) and are not case sensitive.*

2.   Append the path using code.

You can also temporarily add a path to your script. For example, if the scripts you want to call are in the folder C:\Sharedscripts, you can use the following code prior to calling the function:

```
sys.path.append("C:/Sharedscripts")
```

**Note:** *Because this is Python code, you need to use a forward slash (/) for the path.*

The `sys.path.append` statement is a temporary solution meant just so a particular script can call a function in another script.

**Note:** *A third alternative is to modify the PYTHONPATH variable directly from within the operating system. However, this is somewhat cumbersome and error-prone, and therefore not recommended.*

# 12.4 Organizing code into modules

By creating a script that defines a custom function, you are turning the script into a module. All Python script files are, in fact, modules. That's why you can call the function by first importing the script (module), and then using a statement such as `<module>.<function>`. Recall the example:

```
import random
x = random.randint(1,100)
print x
```

The `random` module consists of the random.py file and is located in one of the folders that Python automatically recognizes, C:\Python27\ArcGIS10.1\lib. The random.py script (module) contains a number of functions, including `randint`.

This makes it easy to create new functions in a script and call them from another script. However, it also introduces a complication: how do you distinguish between running a script by itself and calling it from another script? What is needed is a structure that provides control of the execution of the script. If the script is run by itself, the function is executed. If the module is imported into another script, the function is not executed until it is specifically called.

Consider the example hello.py script, which contains a function as well as some test code to make sure the function works:

```
def printmessage():
    print "Hello world"
print message()
```

This type of testing is reasonable, because when you run the script by itself, it confirms that the function works. However, when you import this module to use the function, the test code runs, as follows:

```
>>> import hello
"Hello world"
```

When you import the script file as a module, you don't want the test code to run automatically, but only when you call the specific function. You want to be able to differentiate between running the script by itself and importing it as a module into another script. This is where the variable __name__ comes in (there are two underscores on each side). For a script, the variable has the value of "__main__". For an imported module, the variable is set to the name of the module. Using an `if` statement in the script that contains the function will make it possible to distinguish between a script and a module, as follows:

```
def printmessage():
    print 'Hello world'
if __name__ == '__main__':
    printmessage()
```

In this case, the test of the module will be run only if the script is run by itself. If you import the script, no code will be run until you call the function.

This structure is not limited to testing. In some geoprocessing scripts, almost the entire script consists of one function or more, and only the very last lines of code actually call the function if, indeed, the script is run by itself. The structure is as follows:

```
import arcpy
import os
def mycooltool(<arguments>):
    <line of code>
    <line of code>
    ...
if __name__ == '__main__':
    mycooltool(<arguments>)
```

This structure provides control of the running of the script and makes it possible to use a script in two different ways—running it by itself or calling it from another script.

Consider the earlier example of the `random` module. The very last lines of code look like the example in the figure.

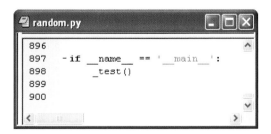

If the random.py script is run by itself, it will run the test function, as shown in the figure.

```
848
849    - def _test(N=2000):
850          _test_generator(N, random, ())
851          _test_generator(N, normalvariate, (0.0, 1.0))
852          _test_generator(N, lognormvariate, (0.0, 1.0))
853          _test_generator(N, vonmisesvariate, (0.0, 1.0))
854          _test_generator(N, gammavariate, (0.01, 1.0))
855          _test_generator(N, gammavariate, (0.1, 1.0))
856          _test_generator(N, gammavariate, (0.1, 2.0))
857          _test_generator(N, gammavariate, (0.5, 1.0))
858          _test_generator(N, gammavariate, (0.9, 1.0))
859          _test_generator(N, gammavariate, (1.0, 1.0))
860          _test_generator(N, gammavariate, (2.0, 1.0))
861          _test_generator(N, gammavariate, (20.0, 1.0))
862          _test_generator(N, gammavariate, (200.0, 1.0))
863          _test_generator(N, gauss, (0.0, 1.0))
864          _test_generator(N, betavariate, (3.0, 3.0))
865          _test_generator(N, triangular, (0.0, 1.0, 1.0/3.0))
866
```

Running the script produces output than can be examined to ensure the random function performs as expected. The output from running the random.py script is as follows:

```
2000 times random
0.0 sec, avg 0.490386, stddev 0.290092, min 0.000360523, max 0.999743
2000 times normalvariate
0.015 sec, avg -0.0379325, stddev 1.01517, min -3.31413, max 3.54333
2000 times lognormvariate
0.0 sec, avg 1.55066, stddev 1.96947, min 0.0308862, max 24.7307
```

These results are printed only when the random.py script is run by itself and not when it is imported as a module into another script.

# 12.5 Using classes

In the previous sections, you saw how to create your own functions and organize your code into modules. This substantially increases code reusability because you can write a section of code and use it many times by calling it from within the same script or from another script. However, these functions and modules have their limitations. The principal limitation is that a function does not store information the way a variable does. Every time a function is run, it starts from scratch.

In some cases, functions and variables are very closely related. For example, consider a land parcel with a number of variables, such as the land-use type, total assessed value, and total area. The parcel may also have procedures associated with it, such as how to estimate the property taxes based on land-use type and total assessed value. These functions require the value of the variables. These values can be passed to a function as arguments. What if a function needs to change the variables? The values could be returned by the function. However, the passing and returning of variables can become quite cumbersome.

A better solution is to use a class. A class provides a way to group together functions and variables that are closely related so they can interact with each other. A class also makes it possible to work with multiple objects of the same type. For example, each land parcel is likely to have the same attributes. The concept of grouping together functions and variables related to a particular type of data is called object-oriented programming (OOP). Classes are the container for these related functions and variables. Classes make it possible to create objects that have specific properties as defined by these functions and variables.

You have seen several ArcPy classes, such as the `env` class, which can be used to access and set environment settings, and the `Result` class, which defines the properties and methods of result objects that are returned by geoprocessing tools. Being able to create your own classes in Python, however, opens up many new possibilities.

To make a class in Python, you use the keyword `class`. Take a look at a simple example:

```
class Person(object):
    def setname(self, name):
        self.name = name
    def greeting(self):
        print "My name is (0).".format(self.name)
```

The `class` keyword is used to create a Python class called `Person`. The class contains two method definitions—these are like function definitions, except that they are written inside a class statement and are therefore referred to as methods. The `self` parameter refers to the object itself. You can call it whatever you like, but it is almost always called "self" by convention.

A class can be thought of as a blueprint. It describes how to make something and you can create many *instances* from this blueprint. Each object created from a class is called an instance of the class. Creating an instance of a class is sometimes referred to as *instantiating* the class.

Next, you will see how this class can be used.

```
me = Person()
```

> **>>> TIP**
>
> The Style Guide for Python Code recommends using the CapitalizedWords, or CapWords, convention for class names—for example, `MyClass`. By contrast, the recommended style for variables, functions, and scripts is all lowercase.

Using an assignment statement creates an instance of the Person class. Creating this instance looks like calling a function. Once an instance is created, you can use the properties and methods of the class, as follows:

```
me.setname("Abraham Lincoln")
me.greeting()
```

Running this code prints the following:

```
My name is Abraham Lincoln.
```

This example is relatively simple, but it illustrates some key concepts. First, a class can be created using the `class` keyword. Second, properties of the class are defined as *methods*—they look like functions but are called methods when they are defined inside a class. Third, a class can contain multiple properties and methods.

Now return to the example of a parcel of land. You want to create a class called `parcel` that has two properties (land-use type and total assessed value) and a procedure (calculating tax) associated with it. For the purpose of this example, assume the property tax is calculated as follows:

- For single-family residential, tax = 0.05 * value.

- For multifamily residential, tax = 0.04 * value.

- For all other land uses, tax = 0.02 * value.

Creating the `Parcel` class is coded as follows:

```
class Parcel(object):
    def __init__(self, landuse, value):
        self.landuse = landuse
        self.value = value

    def assessment(self):
        if self.landuse == "SFR":
            rate = 0.05
        elif self.landuse == "MFR":
            rate = 0.04
        else:
            rate = 0.02
        assessment = self.value * rate
        return assessment
```

The class called `Parcel` is created using the `class` keyword. The class contains two methods: `__init__` and `assessment`. The `__init__` method is a special method reserved for *initializing* objects inside a

class—that is, constructing objects before they can be used. This method
has three arguments: `self`, `landuse`, and `value`. When the class is called,
however, the first argument (`self`) is not used. The argument `self` represents the object and is provided for implicitly by calling the class. The
`assessment` method is where the actual calculation occurs.

Next, take a look at how to use this class. The following code creates an
instance of the parcel object:

```python
myparcel = Parcel("SFR", 200000)
```

With the instance created, you can use its object properties and methods, as
follows:

```python
print "Land use: ", myparcel.landuse
mytax = myparcel.assessment()
print mytax
```

Running this code prints:

```
Land use: SFR
10000.0
```

You can create multiple instances of this object. In a typical scenario, you
could run the property tax calculation for every parcel in a database, creating a new instance for each parcel.

In many cases, you may want to use the class in more than one script.
This can be accomplished by putting it in a module—that is, creating a separate script with the definition of the class, which can then be called from
another script. This is analogous to creating a separate script for a function,
which can be called from other scripts, as described earlier in this chapter.

In this example, the script containing the class is called parcelclass.py
and is as follows:

```python
class Parcel(object):
    def __init__(self, landuse, value):
        self.landuse = landuse
        self.value = value

    def assessment(self):
        if self.landuse == "SFR":
            rate = 0.05
        elif self.landuse == "MFR":
            rate = 0.04
        else:
            rate = 0.02
        assessment = self.value * rate
        return assessment
```

In this example, the script that uses the class is called parceltax.py and is as follows:

```
import parcelclass
myparcel = parcelclass.parcel("SFR", 200000)
print "Land use: ", myparcel.landuse
mytax = myparcel.assessment()
print mytax
```

# 12.6 **Working with packages**

When you have a number of different functions and classes, it often makes sense to put them in separate modules (scripts). As your collection of modules grows, you can consider grouping them into packages. A package is essentially another type of module, but it can contain other modules as well. A regular module is stored as a .py file, but a package is a folder (or directory). Technically speaking, a package is a folder with a file called "__init__.py" in it. This file defines the attributes and methods of the package. It doesn't actually need to define anything; it can just be an empty file, but it must exist. If __init__.py does not exist, the directory is just a directory, and not a package, and it can't be imported. The __init__.py file makes it possible to import a package as a module. For example, to import ArcPy, you use the `import arcpy` statement, but there is no script file called "arcpy.py." However, there is an arcpy folder with a file called "__init__.py."

For example, if you had a package you wanted to call "mytools," you would need to have a folder called "mytools" and inside this folder would need to be a file called "__init__.py". The structure of a package called mytools with two modules (`analysis` and `model`) would look as follows:

`~/Python`—a directory in PYTHONPATH

`~/Python/mytools`—a directory for the mytools package

`~/Python/mytools/__init__.py`—package code

`~/Python/mytools/analysis.py`—analysis module

`~/Python/mytools/model.py`—model module

To use the package, your code would look as follows:

```
import mytools
output = mytools.analysis.<function>(<arguments>)
```

You may wonder what a site package is. A site package is a locally installed package that is available to all users using that computer. The "site" is the local computer. What makes a package a site package has to do with how it is installed, and not its actual contents. During the installation of a site package, the path to the package is added to the PYTHONPATH variable. As a result, the package can be directly imported without first having to add the path.

Python has a number of built-in site packages, which can be found in the Lib\site-packages folder. You will see PythonWin listed there. Parts of the PythonWin editor are installed as a site package, although the actual application is a file called PythonWin.exe, which is located outside the package. Another commonly used site package is NumPy, which is used to manipulate large arrays of data.

ArcPy is referred to as a site package because a typical ArcGIS installation includes both ArcPy and Python, and the folder where ArcPy is located is automatically recognized by Python through the Desktop10.1.pth file located in the Lib\site-packages folder. Where exactly is ArcPy installed? Typically, the location is C:\Program Files\ArcGIS\Desktop10.1\arcpy.

**Note:** *Although the preceding path is the default location for the installation of ArcGIS, it can vary depending on the operating system and the user-defined selections during installation. However, if you can find the ArcGIS installation folder, you will also be able to find the arcpy folder.*

When you explore the contents of this folder, you will find a subfolder called arcpy (which gives ArcPy its name), as shown in the figure, and which contains a file called __init__.py, which makes it a Python package, in addition to many files whose names sound familiar (analysis.py, cartography.py, geocoding.py, and more).

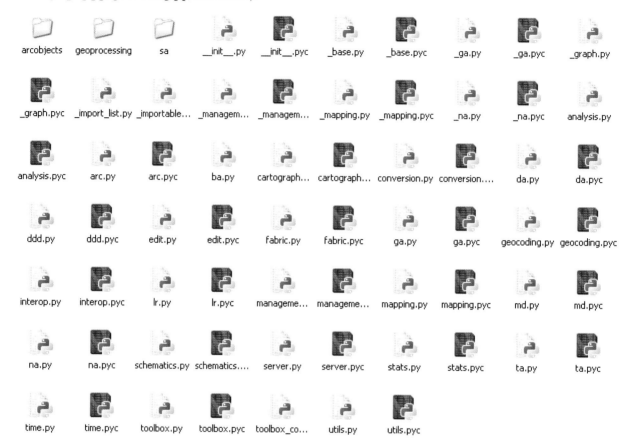

Normally, you should never work with these files directly, but for educational purposes, it is OK to examine them. Just don't make any changes! As part of the installation of ArcPy, the path C:\Program Files\ArcGIS\Desktop10.1\arcpy is added to the PYTHONPATH environment variable in Windows and you can start using ArcPy immediately.

# Points to remember

- Custom functions can be created using the `def` statement. The block of code that follows the `def` statement defines what the function actually does. Custom functions can contain arguments, although they are not required.

- Custom functions can be called from within the same script or from another script. When calling a function from another script, you import the script that contains the function as a module. A custom module is therefore a regular .py file that contains at least one custom function.

- To distinguish between running a script by itself and importing it as a module into another script, you can use the `if __name__ == 'main__':` statement.

- When importing modules, you cannot use paths, and modules (scripts) need to be located in the same folder as the script importing the module(s) or in the folders included in the PYTHONPATH environment variable. As needed, you can permanently add a path by using a .pth file in the site-packages directory or temporarily add a path in your script using the `sys.path.append` statement.

- Custom classes can be created to make it easier to group together functions and variables related to a particular item. Classes can be called from within the same script or from another script.

- As your collection of custom functions and classes grows, you can consider making it a package, similar to the ArcPy site package.

# Part 4

**Creating and using
script tools**

# Chapter 13
## Creating custom tools

## 13.1 Introduction

This chapter describes the process of turning a Python script into a tool. Tools make it possible to integrate your scripts in ArcGIS. Tools can be run from ArcToolbox, can be used within a model, and can be called by other scripts. Tools have a tool dialog box, which typically contains the parameters that are passed to the script. Developing tools is relatively easy and greatly enhances the experience of using a script. Tool dialog boxes reduce user error because parameters can be specified using drop-down lists, check boxes, combo boxes, and other mechanisms. This provides substantial control of user input, greatly reducing the need to write a lot of error-checking code. Creating tools also makes it easier to share scripts with others.

## 13.2 Why create your own tools?

Many ArcGIS workflows consist of a sequence of operations in which the output of one tool becomes the input of another tool. ModelBuilder and scripting can be used to automatically run these tools in a sequence. Any model created and saved using ModelBuilder is a tool because it is located in a toolbox (.tbx file) or a geodatabase. A model, therefore, is always run from within an ArcGIS for Desktop application, such as ArcMap or ArcCatalog. A Python script (.py file), however, can be run in two ways:

1. *As a stand-alone script.* This means the script is run from the operating system or from within a Python editor, such as PythonWin. For a script to use geoprocessing tools, ArcGIS for Desktop needs to be installed and licensed, but no ArcGIS for Desktop application needs to be open for the script to run. For example, you can schedule a script to run at a prescribed time directly from the operating system.

2. *As a tool within ArcGIS.* This means the script is turned into a tool to be run from within an ArcGIS for Desktop application. Such a tool is like any other tool: it is located in a toolbox, can be run from a tool dialog box, and can be called from other scripts, models, and tools.

There are a number of advantages to using tools instead of stand-alone scripts:

- A tool includes a tool dialog box, which makes it easier for users to enter the parameters using built-in validation and error checking.

- A tool becomes an integral part of geoprocessing. This makes it possible to access the tool from the Catalog and Search windows in ArcMap. It also makes it possible to use the tool in ModelBuilder and in the Python window, and to call it from another script.

- A tool is fully integrated with the application it was called from. This means any environment settings are passed from the application, such as ArcMap, to the tool.

- The use of tools makes it possible to write messages to the Results window.

- Documentation can be provided for tools, which can be accessed like the documentation for system tools.

- Sharing a tool makes it easier to share the functionality of a script with others.

- A well-designed tool means a user requires no knowledge of Python to use the tool's functionality.

# 13.3  **Steps to creating a tool**

A tool is created using the following steps:

1. Create a Python script and save it as a .py file.

2. Create a custom toolbox (.tbx file) where the tool can be stored.

3. Add a tool to the custom toolbox using the Add Script wizard.

4. Modify the script with input and output variables so that it is seam-lessly integrated into the geoprocessing framework.

You can create a new custom toolbox in ArcCatalog or in the Catalog window inside another ArcGIS for Desktop application. Navigate to Toolboxes, right-click My Toolboxes, and click New > Toolbox. Give the toolbox a name.

> **Note:** *Do not click New > Python Toolbox, because a Python toolbox is created entirely in Python and not in ArcGIS. A new generic toolbox is all that is needed at this point.*

This section describes how to create a script tool using a custom toolbox. Although ArcGIS 10.1 has introduced Python toolboxes, which support additional capabilities to custom toolboxes, it is more convenient to use a custom toolbox when starting to create your first script tools.

Your empty custom toolbox can now be added to ArcToolbox. You can drag it from the Catalog window into ArcToolbox or right-click inside ArcToolbox and click Add Toolbox, which allows you to browse for a tool-box in any folder.

A toolbox, which consists of a single .tbx file, can be located anywhere on your computer. The folder My Toolboxes is one logical location to organize custom toolboxes, but they can also be located in any folder where datasets and other files for a particular project are organized—for example, C:\EsriPress\Python\Data\MyCoolTools.tbx. Custom toolboxes can also be located inside a geodatabase. Like other elements, a toolbox inside a geodatabase no longer has a file extension—for example, C:\EsriPress\Python\Data\study.gdb\MyCoolTools.

To create a tool, in ArcToolbox, right-click a custom toolbox and click Add > Script. Write access to the toolbox is needed to be able to add a new tool. As a result, you cannot add tools to any of the system toolboxes in ArcToolbox.

The Add Script wizard has three panels. The first panel is used to specify the script name, label, and description. The second panel is used to specify the actual script file (.py), including its path. The third panel is used to specify the tool's parameters. Each of these panels is reviewed here in detail.

The example script that comes next illustrates how to create a tool. This script creates a list of all the feature classes in a workspace and copies these feature classes to an existing file geodatabase, as follows:

```python
# Python script: copyfeatures.py
# This script copies all feature classes from a workspace into
# a file geodatabase.

# Import the ArcPy package.
import arcpy
import os

# Set the current workspace.
from arcpy import env
env.workspace = "C:/Data"

# Create a list of feature classes in the current workspace.
fclist = arcpy.ListFeatureClasses()

# Copy each feature class to a file geodatabase - keep the same
# name but use the basename property to remove any file
# extensions, including .shp.
for fc in fclist:
    fcdesc = arcpy.Describe(fc)
    arcpy.CopyFeatures_management(fc, os.path.join("C:/Data/study.gdb/",  ➤
➤ fcdesc.basename))
```

This script is written as a stand-alone script. Both the current workspace and the file geodatabase are hard-coded in the script. Although the script will run correctly, it will require modification to be useful as a tool.

To start the Add Script wizard, in ArcToolbox, right-click a custom toolbox and click Add > Script. This brings up the first panel of the wizard, as shown in the figure. ➜

The first panel of the wizard is used to specify the tool name, label, description, and style sheet as follows:

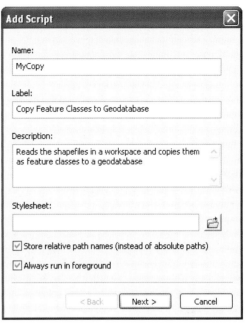

- The name of a tool is used when you want to run a tool from Python. The name cannot contain any spaces.

- The label of the tool is the display name in ArcToolbox. The label name can have spaces.

Consider the example of the Get Count tool. In ArcToolbox, the tool appears with its label, Get Count (with a space), but for the tool to be called from Python, its name, GetCount (without a space), is used.

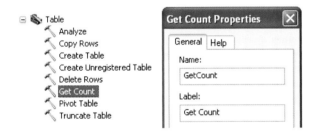

- The description is an optional field to provide a customized description. The text is automatically used to provide the contents of the Help panel on a tool dialog box.

- An optional style sheet can be selected. If none is selected, the default style sheet is used. Style sheets are used to control the properties of items on a tool dialog box. A style sheet provides style and layout information. All the system tools in ArcToolbox use the default style sheet. Typically, you want your custom tools to look just like the system tools so the default style sheet is usually sufficient.

- Optionally, the "Store relative path names" check box can be selected. When it is selected, relative paths are used instead of absolute paths to reference the location of the script file in relation to the location of the custom toolbox (.tbx) file. Only the path to the script file can be stored as a relative path; paths within the script itself will not be converted. If you are going to share the tool with others, it is a good idea to select this check box.

- Optionally, the "Always run in foreground" check box can be selected. This will ensure the script is run using foreground processing, even if background processing has been enabled under Geoprocessing Options. (Background processing allows you to continue to work in the ArcGIS for Desktop application while the tool is running.) Some scripts require foreground processing—for example, mapping scripts, which use the CURRENT keyword to obtain the active map document in ArcMap. For other scripts, selecting foreground or background processing is a matter of preference.

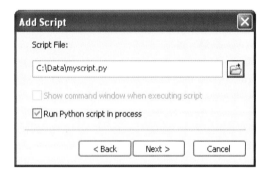

In the second panel of the Add Script wizard, you can set the following:

- The complete path of the Python script file to be run. You can browse to an existing file or type the path of a file. If you specify a script file that does not exist, you will be prompted to create a new (empty) script file. Alternatively, the script file field can be left blank and added later.

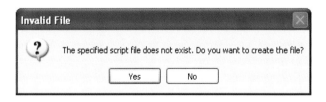

- The "Show command window when executing script" check box, which is cleared by default. When the box is selected, an additional window appears during tool execution to show messages that are not part of the regular geoprocessing messages but are written to the standard output for Python. For example, the Python `print` statement writes to the standard output, such as the Interactive Window in PythonWin. If such statements are part of your script, the messages would not appear unless this box is selected. Scripts that are referenced by a tool should normally use geoprocessing messaging and not write to the standard output. So the box remains clear unless you have very specific needs for viewing messages.

- The "Run Python script in process" check box, which is selected by default. Python scripts run faster if they are run "in process," so typically you'll want this option selected. Running in process requires that Python modules in your script be designed to run in process, which is the case for standard modules such as `os`, `math`, and `string`. Nonstandard modules from third parties may not be designed for this process, which can result in performance issues. So if you are using third-party modules in your script, which appears to result in unexpected problems, you can try running the script out of process instead.

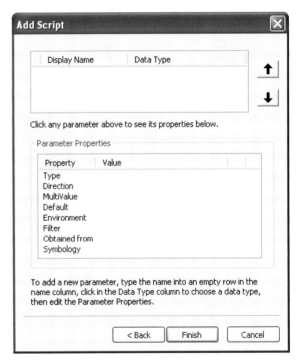

The third panel of the Add Script wizard is used to specify the tool parameters. By default, no parameters are listed, but most tools need at least one input parameter and one output parameter. The top half of the panel allows you to create parameters and the bottom half allows you to set the properties for each of these parameters. Setting parameters is covered in detail later in this chapter, including parameter properties. ➜

To complete the Add Script wizard, click the Finish button. Completing this wizard adds a tool to your custom toolbox.

All the settings in the Add Script wizard can be modified by right-clicking the tool and clicking Properties. The Properties dialog box of the tool includes tabs for General, Source, and Parameters, which correspond to the three panels of the Add Script wizard. Two additional tabs on the dialog box are Validation and Help. These tabs are revisited later in this chapter.

Your new tool can be accessed just like a regular tool. Right-click the tool and click Open or simply double-click the tool in its toolbox. The tool dialog box seems empty at this point because no parameters were set in the third panel of the Add Script wizard. The Help panel on the right shows the tool's description, but there are no parameters yet.

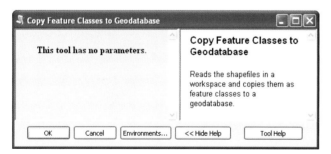

Clicking OK runs the tool—that is, runs the script—but without parameters, it does not provide the user with any control over its execution. One of the most critical steps in creating tools is to create input and output parameters and add them to the tool's dialog box. Remember, however, that the script was written to run as a stand-alone tool. Setting parameters therefore requires modifying the code in the script so it can receive the parameters set by the tool dialog box.

# 13.4 Editing tool code

When you create a tool, you typically have to make changes to the script so that the tool dialog box and the script can interact seamlessly. When testing a tool, you will alternate between running the tool and editing the script until the tool works as desired. You can leave the Python editor open while you do it. You can open a script from within the Python editor, but there is also a shortcut in ArcGIS. Right-click the tool in the toolbox and click Edit. This will open the Python script in a Python editor such as IDLE or PythonWin. The Python editor that is used is determined by the geoprocessing options. To change these settings, on the menu bar in an ArcGIS for Desktop application, click Geoprocessing > Geoprocessing Options. ➔

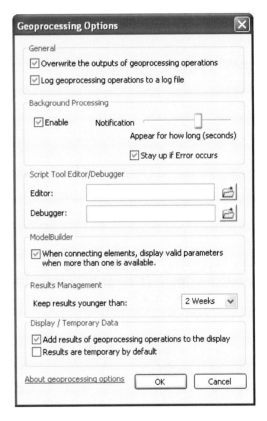

To select PythonWin as the editor, browse to the location of the PythonWin application. Typically, this application is located at C:\Python27\ArcGIS10.1\Lib\site-packages\PythonWin\PythonWin.exe, but this may vary depending on how Python was installed on your computer. Once you set a specific editor, any script file opened from within ArcGIS will open in this editor.

Now that a tool is created and you can access its code, it is time to take a closer look at parameters.

# 13.5 Exploring tool parameters

As you have seen in several earlier chapters, all geo-processing tools have parameters. Parameter values are set on the tool dialog box. In a stand-alone script, parameters are typically set within the script unless user input is expected. For tools, parameters can be set using the tool dialog box. When a tool runs, the parameter values from the tool dialog box are passed to the script. The script reads these values and uses them in the code. Creating and exposing parameters requires the following steps:

- Including code in the script to receive the parameter values

- Setting up the parameters in the tool's properties

Next, you will try this out by using one of the built-in tools, the Multiple Ring Buffer tool. The tool dialog box is shown in the figure. ➔

The Multiple Ring Buffer tool has seven parameters total, three of which are optional. The Parameters tab on the tool properties dialog box lists the same seven parameters, in the same order, as the tool dialog box. It also shows the data type of each parameter. For example, the input features consist of a feature layer, and the buffer unit is a string value. ➔

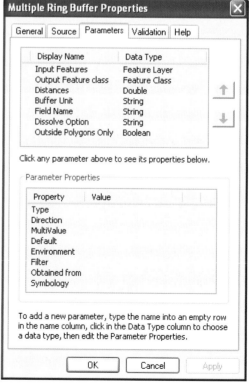

> *Note: Because the Multiple Ring Buffer tool is a built-in tool, you can see the list of parameters, but you cannot see or edit the parameter properties. And because the parameters are read-only, the entire panel appears dimmed. If you want to read more about the parameters, you can copy the tool to a custom toolbox, which provides read/write permission to access its properties.*

Once a user specifies the parameter values on the Multiple Ring Buffer
tool dialog box, the tool can be run. Once the tool is run, the user-specified
parameter values will be passed to the script. Take a look at the script's
code to see how these parameter values are received by the script. The
example MultiRingBuffer.py script, which is shown in the PythonWin
editor in the figure, includes the import of a number of modules and some
introductory comments, and then contains a section of code where the
parameter values are received.

```python
import arcgisscripting
import os
import sys
import types
import locale

gp = arcgisscripting.create(9.3)

#Define message constants so they may be translated easily
msgBuffRings   = gp.GetIDMessage(86149) #"Buffering distance "
msgMergeRings  = gp.GetIDMessage(86150) #"Merging rings..."
msgDissolve    = gp.GetIDMessage(86151) #"Dissolving overlapping boundaries..."

def initiateMultiBuffer():

    # Get the input argument values
    # Input FC
    input          = gp.GetParameterAsText(0)
    # Output FC
    output         = gp.GetParameterAsText(1)
    # Distances
    distances      = gp.GetParameter(2)
    # Unit
    unit           = gp.GetParameterAsText(3)
    if unit.lower() == "default":
        unit = ""
    # If no field name is specified, use the name "distance" by default
    fieldName      = checkFieldName(gp, gp.GetParameterAsText(4), os.path.dirname(output))
    #Dissolve option
    dissolveOption = gp.GetParameterAsText(5)
    # Outside Polygons
    outsidePolygons = gp.GetParameterAsText(6)
    if outsidePolygons.lower() == "true":
        sideType = "OUTSIDE_ONLY"
    else:
        sideType = ""

    createMultiBuffers(gp, input, output, distances, unit, fieldName, dissolveOption, sideType)
```

The tool's parameters are received by the script using the GetParameterAsText and the GetParameter functions. This script uses the ArcGISscripting module from version 9.3, but the concept is the same. The syntax of the GetParameterAsText function is

```
<variable> = arcpy.GetParameterAsText(<index>)
```

The only argument of this function is an index number on the tool's dialog box, which indicates the numeric position of the parameter. The parameters set on the tool dialog box are sent to the script as a list and the GetParameterAsText function assigns these parameter values to variables in the script. The two figures (see below and facing page) show how each parameter on the tool dialog box matches the code index number—for example, Input Features is (0), Output Feature class is (1), and Distances is (2).

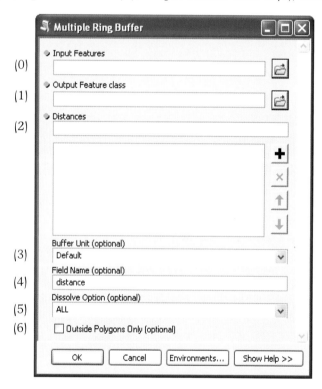

```
def initiateMultiBuffer():

    # Get the input argument values
    # Input FC
    input           = gp.GetParameterAsText(0)
    # Output FC
    output          = gp.GetParameterAsText(1)
    # Distances
    distances       = gp.GetParameter(2)
    # Unit
    unit            = gp.GetParameterAsText(3)
    if unit.lower() == "default":
        unit = ""
    # If no field name is specified, use the name "distance" by default
    fieldName       = checkFieldName(gp, gp.GetParameterAsText(4), os.path.dirname(output))
    #Dissolve option
    dissolveOption  = gp.GetParameterAsText(5)
    # Outside Polygons
    outsidePolygons = gp.GetParameterAsText(6)
    if outsidePolygons.lower() == "true":
        sideType = "OUTSIDE_ONLY"
    else:
        sideType = ""

    createMultiBuffers(gp, input, output, distances, unit, fieldName, dissolveOption, sideType)
```

The GetParameterAsText function receives parameters as a text string,
even if the parameter on the tool dialog box is a different data type. Numerical values, Boolean values, and other data types are all converted to strings
and additional code is included to correctly interpret these strings. For
example, the code of the Outside Polygons Only parameter is as follows:

```
outsidePolygons = gp.GetParameterAsText(6)
if outsidePolygons.lower() == "true":
    sideType = "OUTSIDE_ONLY"
else:
    sideType = ""
```

The Outside Polygons Only parameter on the tool dialog box is a Boolean
value of True or False. These values are converted to strings, and as a result,
the conditional statement uses the string value "true" instead of the Boolean value True.

The Distances parameter on the tool dialog box is received by the script using the `GetParameter` function. This is because the parameter consists of a list of values (doubles) instead of a single value. The `GetParameter` function reads this list as an object.

> **Note:** *An alternative to using the* `GetParameterAsText` *and* `GetParameter` *functions is to use* `sys.argv`, *or system arguments. The use of* `sys.argv` *has certain limitations, including that it accepts a limited number of characters. Using the* `GetParameterAsText` *and* `GetParameter` *functions is therefore preferred. Prior to ArcGIS 9.2, these functions would work only for tools referencing a script, and stand-alone scripts could use only* `sys.argv`. *The latter is therefore relatively common in older scripts. The index number for* `sys.argv` *starts at 1, so* `sys.argv[1]` *is equivalent to* `GetParameterAsText(0)`.

Every tool parameter has an associated data type. One of the benefits of data types is that the tool dialog box will not send values to the script unless they are the correct data type. User entries for parameters are checked against the parameter data types before they are sent to the script. This is one advantage of using tools over stand-alone scripts because the script does not have to check for invalid parameters.

The data types of the parameters of the Multiple Ring Buffer tool include a feature layer, a feature class, a double, three strings, and a Boolean. Many more data types are possible for the parameters of a custom tool, from an address locator to a Z domain. Data types of parameters should be selected carefully because they control the interaction between the tool dialog box and the script. ➔

After parameters are assigned a data type, the tool dialog box uses this

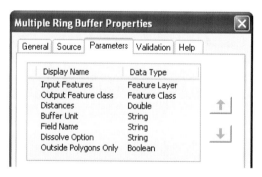

data type to check the parameter value. For example, if you enter a path to an element of a different data type, the tool dialog box will generate an error. In the example in the two figures, the Input Features parameter is a feature layer (upper right), so typing the path for a raster, such as C:\Data\dem, will generate an error (lower right) and prevent the tool from running. This built-in error-checking mechanism prevents users from using incorrect parameters to run a tool. When the tool runs, the dialog box has already validated the parameter Input Features as a feature layer, and no additional code is needed in the script to verify it.

The data type property is also used to browse through folders for data. Only data that matches the parameter's data type will be shown. This prevents the entering of incorrect paths to data.

# 13.6 **Setting tool parameters**

Tool parameters can be set in the Add Script wizard when creating the tool. They can also be edited after the tool has been created by accessing the tool's properties dialog box. Setting parameters is the same, no matter which method is used.

A parameter is added by placing the cursor in the first empty cell in the Display Name column, under the Parameters tab, and typing a name for the parameter, which is displayed on the tool dialog box. Next, the data type is specified by selecting from an extensive drop-down list. ➔

You can add more parameters by repeating these steps. Once multiple parameters are created, the order can be changed by selecting one and using the arrow keys to move the row up or down.

Each parameter has a number of properties, as shown in the bottom half of the dialog box. When each parameter is created, its properties are set to default values based on the parameter's data type. Some of the key parameters are discussed as follows. A complete description of all the parameters can be found in ArcGIS Desktop Help, under the topic "Setting script tool parameters."

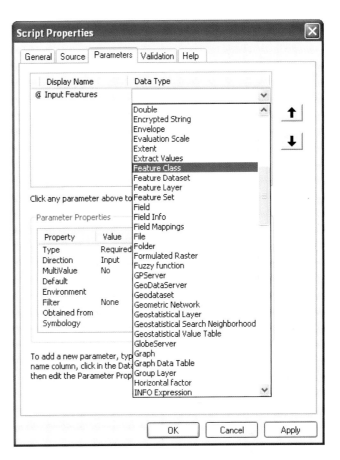

## Type

There are three choices for Type: Required, Optional, and Derived. Required means that a parameter value needs to be specified for a tool to run. Optional means that a value is not required for a script to run. Typically, this means that a default value is specified. Derived parameters are used for output parameters only and do not appear on the tool dialog box.

Derived parameters are used in a number of cases, including the following:

• When a tool outputs a single value instead of a dataset. Such a single value is often referred to as a *scalar*.

• When a tool creates outputs using information from other parameters.

• When a tool modifies an input without creating a new output.

All tools should have outputs so that the tool can be used in a model. Sometimes the only way to accomplish this is by using derived parameters. Examples of tools with derived parameters include the Get Count tool and the Add Field tool.

The input parameter of the Get Count tool is a feature class, table, layer, or raster, and the output is a count of the number of rows. This count is a scalar variable and is returned as a result object. It comprises an output parameter and is a derived parameter that does not appear on the tool dialog box.

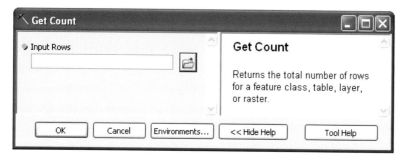

The Add Field tool adds a new field to an input table. The input table is a required parameter, and so is the name of the new field. The rest of the parameters are optional, as shown in the figure.

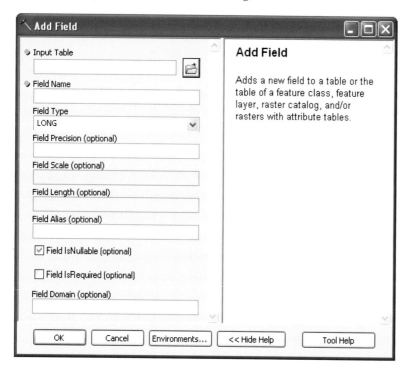

*Note: Running the Get Count tool as a single tool from ArcToolbox is not common. Although the count is printed to the Results window, this tool is typically used within a model or a script where the output is used as the input to another step. The Get Count tool is also commonly used in conditional statements. For example, a particular procedure can be stopped if the count of rows is zero (0).*

When the tool runs, a new field is added based on the input parameters. The output of the tool is the modified table or feature class. Because this table or feature class is an input parameter, there is no need to specify the output on the tool dialog box. The output of the tool is therefore specified as a derived parameter. An easy way to visualize it is by using the tool in a model, as shown in the figure.

When you inspect the properties of the input feature class and the output, you will notice that they reference the exact same feature class. In fact, you cannot specify or change the name of the output (you can change the display name in the model but not the underlying data).

### Direction

The Direction property defines whether the parameter is an input of the tool or an output of the tool. For derived parameters, the parameter direction will automatically be set to output. Every tool should have output parameters. This makes it possible to use the tool in ModelBuilder. Although technically a script can run without output parameters, for ModelBuilder to work, every tool needs an output so it can be used as the input to another tool in the model.

### MultiValue

Some tool parameters consist of a list of values rather than a single value. When the MultiValue property is set to No, only a single value can be used. When the MultiValue property is set to Yes, a list of values can be used.

Multivalue parameters are quite common for built-in geoprocessing tools. For example, the Union tool uses a list of input feature classes. The Union tool uses the default multivalue parameter control, which is simply a list of inputs that can be added, removed, and reordered. ➔

The second type of multivalue parameter is a list of check boxes. This is commonly used for fields, as illustrated by the Delete Field tool. Check boxes can also be used if a Value List filter is applied, which is discussed later in this section.

Multivalue parameters are passed to the script as a delimited string, with the individual list elements separated by semicolons (;). The Python `split` method can be used to create a list of the elements from the string. The syntax is as follows:

```
import arcpy
input = arcpy.GetParametersAsText(0)
inputList = input.split(";")
```

As an alternative, you can use `GetParameter` to obtain a ValueTable object instead of a string. In a ValueTable, the values are stored in a virtual table of rows and columns. ValueTable objects are specially designed for multi-value parameters.

This means that when writing the script, you need to be aware of the parameter types being passed to the script from the tool dialog box.

### Default

The default value of a parameter is the contents of the parameter when the script tool's dialog box is open. If no default value is specified, the parameter value will be blank on the tool's dialog box. If a default value is specified, the Environment property will be disabled.

### Environment

A default value for a parameter can also be specified using the Environment property. By right-clicking the cell next to Environment, you can choose the name of the environment setting. Once this property is set, the default value is obtained from the environment settings of the geoprocessing framework. If the Environment property is specified, the default property will be ignored—so you need to specify one or the other.

### Filter

The Filter property allows you to limit the values of dataset types to be entered for a parameter. There are a number of filter types, and the type depends on the data type of the parameter. The different filter types are Value List, Range, Feature Class, File, Field, and Workspace.

For most data types, there is only one filter type. For example, if the data type of a parameter is set to Feature Class, the only possible filter type is Feature Class. The only exceptions are the Long and Doubles data types, which have Value List and Range as possible filter types. Many data types have no filter type at all.

The different filter types provide specific control of which parameters are valid inputs. Carefully setting the filter type will improve the robustness of the tool. The different filter types are discussed in more detail in ArcGIS Desktop Help, under the topic "Setting script tool parameters."

## Obtained from

In many cases, a tool parameter is closely related to another one. For example, consider the Delete Field tool. ➔

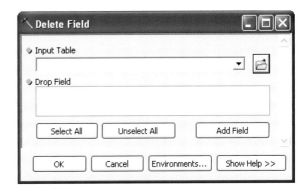

The first parameter is an input table, and the second parameter, Drop Field, is a list of fields. The list of fields is populated only when the input table is selected, as shown in the figure. ➔

This dependency of a parameter on another parameter in the same tool is controlled using the "Obtained from" property. In the example of the Delete Field tool, the Obtained from property of the Drop Field parameter is set to the input table.

A second reason to use the Obtained from property is to work with derived output parameters. For example, when an input parameter is modified by a tool, the Obtained from property of the derived output parameter is set to the input parameter. In the case of the Delete Field tool, the Obtained from property of the output parameter is set to the input table.

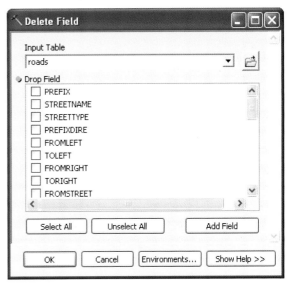

*Note: Remember that the derived output parameter is not visible on the tool dialog box.*

## Symbology

By default, the output of a geoprocessing tool is added to the ArcMap table of contents. This behavior can be set on the Geoprocessing Options dialog box under Display/Temporary Data.

The symbology of a layer added in this way follows the same rules as when data is added using the Add Data option in ArcMap—in other words, there is no customized symbology. The Symbology property can be set to a custom layer file (.lyr). This option is available only for outputs where layer files make sense, such as feature classes, rasters, TINs, and the like. The parameter type can be required or derived, but the parameter direction has to be set to output for the Symbology property to be accessible. ➔

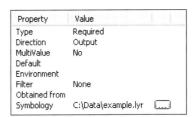

Notice that setting the Symbology property does not control whether the output will be added to the table of contents, because this is controlled by Geoprocessing Options in ArcMap.

## 13.7 Examining an example script tool

The following example illustrates the steps to convert a stand-alone script to a script tool. The following stand-alone script was introduced at the beginning of the chapter. The script creates a list of all the shapefiles in a workspace and copies them to a geodatabase. For the purpose of this example, the script is located in the C:\Sharedscripts folder. The script is as follows:

```
# Python script: copyfeatures.py
# This script copies all feature classes from a workspace to
# a file geodatabase.

# Import the ArcPy package.
import arcpy
import os

# Set the current workspace.
from arcpy import env
env.workspace = "C:/Data"

# Create a list of feature classes in the current workspace
fclist = arcpy.ListFeatureClasses()

# Copy each feature class to a file geodatabase - keep the same
# name but use the basename property to remove any file
# extensions, including .shp
for fc in fclist:
    fcdesc = arcpy.Describe(fc)
    arcpy.CopyFeatures_management(fc, "C:/Data/study.gdb/" + fcdesc.➜
➜ basename)
```

Two workspaces hard-coded into the script have to be modified to parameters using the `GetParameterAsText` function. The revised script (without comments) is as follows:

```
import arcpy
import os
from arcpy import env
env.workspace = GetParameterAsText(0)
outgdb = GetParameterAsText(1)
fclist = arcpy.ListFeatureClasses()
for fc in fclist:
    fcdesc = arcpy.Describe(fc)
    arcpy.CopyFeatures_management(fc, os.path.join(outgdb, fcdesc.➜
➜ basename))
```

A custom toolbox is created in the My Toolboxes folder. This empty toolbox can be dragged to ArcToolbox. ➔

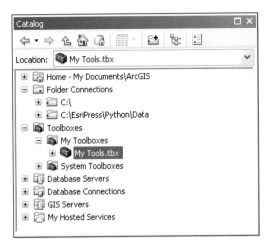

To create a new script tool, right-click the custom toolbox and click Add > Script. In the first panel on the Add Script dialog box, specify the name, label, and description of the script tool. Select the check boxes for storing relative path names and foreground processing. ➔

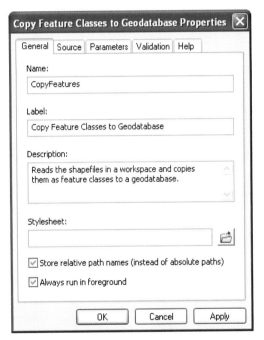

In the second panel on the Add Script wizard, browse to the location of the script file—in this case, C:\Shared-scripts\copyfeatures.py. Leave the other settings to their defaults. ➔

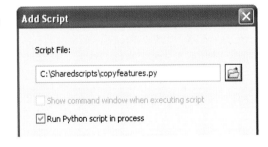

In the third panel on the Add Script wizard, you can specify the parameters. For the first parameter, enter the name **Input Workspace** and select Workspace for Data Type. ➔

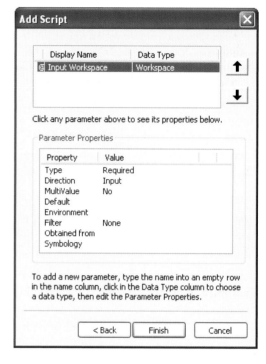

Under Parameter Properties, click the Filter property. Under Value, select Workspace. This brings up the Workspace filter. Under Type, clear the check boxes for "Local database" and "Remote database." ➔

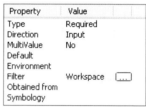

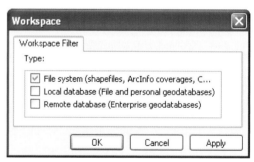

For the second parameter, enter the name **Output Workspace** and select Workspace for Data Type. Under Parameter Properties, set Direction to Output. ➔

In the Add Script wizard, click Finish to complete creating the script tool. After the parameters are added and the properties are set, the tool dialog box should look like the example in the figure. ➜

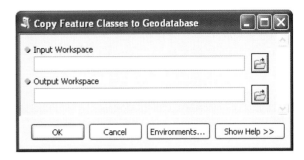

The tool is now ready to run. The tool copies the shapefiles to a geodatabase, and the output of the tool is the geodatabase workspace. The copied feature classes themselves will not be added to the ArcMap table of contents—this is by design because the number of feature classes could conceivably be very large.

# 13.8  Customizing tool behavior

Once a tool's parameters are specified, you can add custom behavior. Examples of custom behavior include the following:

- Certain parameters may need to be enabled or disabled based on the values contained in other parameters.

- Some parameters may benefit from having a default value specified based on the values in other parameters.

- If there are a lot of parameters, it may be more effective to organize parameters into different categories.

- Warning and error messages may need to be customized.

Tool behaviors can be set on the Validation tab on the Script Properties dialog box. In the Validation panel, you can use Python code that uses a Python class called `ToolValidator`. The `ToolValidator` class controls how the tool dialog box is changed based on user input. It is also used to describe the output data the tool produces, which is important for using tools in ModelBuilder.

The `ToolValidator` class was introduced in ArcGIS 9.3, providing more possibilities for creating robust tools. A detailed description of customizing tool behavior is not provided here. Details on the `ToolValidator` class can be found in ArcGIS Desktop Help, under the topic "Customizing script tool behavior."

# 13.9 Working with messages

One of the advantages of running a script as a tool is writing messages that appear on the progress dialog box and in the Results window. Tools and scripts that call a tool also have access to these messages. When scripts are run as stand-alone scripts, messages are printed only to the Interactive Window—there is no progress dialog box and no Results window where messages can be retrieved later. There is also no sharing of messages between stand-alone scripts.

However, because script tools work like any other tool, they automatically print messages to the Results window. For example, when the Copy Feature Classes to Geodatabase tool is run, it prints very simple messages that indicate when the script is running and note when it is completed.

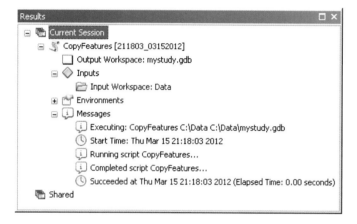

Since this is just the bare bones, there are a number of ArcPy functions for writing messages. These include the following:

- AddMessage—for general information messages (severity = 0)

- AddWarning—for warning messages (severity = 1)

- AddError—for error messages (severity = 2)

- AddIDMessage—for both warning and error messages

- AddReturnMessage—for all messages, independent of severity

The AddReturnMessage function can be used to retrieve all messages returned from a previously run tool, regardless of severity. The original severity of the geoprocessing messages is preserved—for example, an error message is printed as an error message. Some of the other message functions create a custom message. For example, the AddError and

AddMessage functions are used in the following code to print custom messages to the Results window based on the result of a particular tool:

```
import arcpy
fc = arcpy.GetParameterAsText(0)
result = arcpy.GetCount_management(fc)
fcount = int(result.getOutput(0))
if fcount == 0:
    arcpy.AddError(fc + " has no features.")
else:
    arcpy.AddMessage(fc + " has " + str(fcount) + " features.")
```

In the case of a feature class without any features, running this code will produce an error message, as shown in the figure.

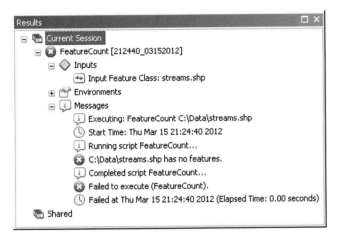

Calling the AddError function also results in a Failed to execute message. However, it does not add an exception, and the code will keep running after the AddError call.

When the AddWarning function is used instead, it results in a warning message, but the script will finish running.

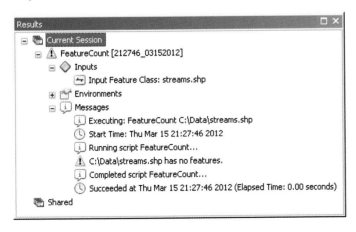

Another level of control can be accomplished using the `AddIDMessage` function. This function makes it possible to use system messages within a tool. The syntax of the function is

```
AddIDMessage(message_type, message_ID, {add_argument1}, {add_argument2})
```

The message type can be set to `Error`, `Informative`, or `Warning`. The message ID number indicates the specific Esri system message. Depending on the message, additional arguments may be necessary. In the following example code, an error message, with the message ID number 12, is produced if the output feature class already exists:

```
import arcpy
infc = arcpy.GetParameterAsText(0)
outfc = arcpy.GetParameterAsText(1)
if arcpy.Exists(outfc):
    arcpy.AddIDMessage("Error", 12, outfc)
else:
    arcpy.CopyFeatures_management(infc, outfc)
```

The syntax of error message 12 is

```
000012 : <value> already exists
```

This message has one argument, which in this case is the name of a feature class. A complete list of tool error and warning messages can be found in ArcGIS Desktop Help, under Geoprocessing > Tool errors and warnings. A small sample of these error messages in Help is shown in the figure.

# 13.10 Handling messages for stand-alone scripts and tools

Python scripts can be run as stand-alone scripts or as tools. Messaging works a bit differently for each one. However, a script can be designed to handle both scenarios.

For a stand-alone script, there is no way to view messages, and they have to be printed to the interactive interpreter. For a tool, functions such as `AddError` are used instead of a print statement to ensure messages appear in the geoprocessing environment, including the Results window.

Standard practice is to write a message-handling routine that writes messages to both the interactive interpreter and the geoprocessing environment, using a `print` statement for the former and ArcPy functions such as `AddError`, `AddWarning`, and `AddMessage` for the latter.

# 13.11 Customizing tool progress information

When a tool runs, information on its progress can take several forms. By default, the geoprocessing framework in ArcGIS uses background processing. This means you can continue to use an application while the geoprocessing operations run in the background. During background processing, a progress bar appears at the bottom of the document on the ArcGIS for Desktop application status bar, as shown in the figure. ➔

When the geoprocessing operation is completed, a pop-up notification appears in the notification area, at the far right of the taskbar. ➔

Background processing can be enabled or disabled on the Geoprocessing Options dialog box. The slider under Background Processing, as shown in the figure, can be moved to control how long the pop-up window appears at the end of background processing. ➔

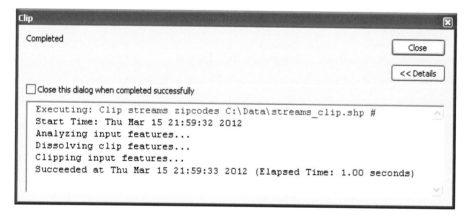

When background processing is disabled, foreground processing is enabled. During foreground processing, a progress dialog box appears. A progress dialog box includes a *progressor*, which consists of a horizontal bar indicating the progress of the tool, and a message area, which consists of a complete list of geoprocessing messages, as shown in the figure. This is the same list of messages that appears in the Results window.

Whether a tool runs as background or foreground processing can be controlled at the application level using the Geoprocessing Options dialog box. For a script tool, background or foreground processing can also be controlled as part of the script tool properties, as described in section 13.3. The appearance of the progress dialog box, which appears during foreground processing, can be controlled using the ArcPy progressor functions. These functions also have an effect on the Results window. The ArcPy progressor functions include the following:

- `SetProgressor`—sets the type of progressor

- `SetProgressorLabel`—changes the label of the progressor

- `SetProgressorPosition`—moves the step progressor by an increment

- `ResetProgressor`—resets the progressor

There are two types of progressors: default and step. In the default type, the progressor moves back and forth continuously but doesn't provide a clear indication of how much progress is being made. The label above the progressor provides information on the current geoprocessing operation.

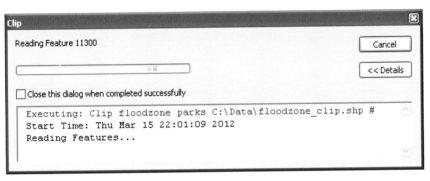

In the step progressor, the percentage completed is shown. This can be useful when processing large datasets.

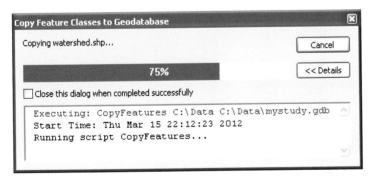

The type of progressor is set using the `SetProgressor` function. This function establishes a progressor object, which allows progress information to be passed to a progress dialog box. The appearance of the progress dialog box can be controlled using either the default progressor or the step progressor.

The syntax of this function is

```
SetProgressor(type, {message}, {min_range}, {max_range}, {step_value})
```

The progressor `type` is either default or step. The `message` is the progressor label that appears at the beginning of the tool's execution. The three remaining parameters are for step progressors only and indicate the start value, the end value, and the step interval. In a typical step progressor, the start value would be set to 0, the end value to however many steps are completed in the geoprocessing operations, and the step interval to 1.

The `SetProgressorLabel` function is used to update the label of the progressor, which is typically a unique string specific to each step. The `SetProgressorPosition` function is used to move the step progressor by an increment based on the percentage of features completed. These functions are commonly used in combination so that the label is updated at every increment.

Once tool execution is completed, the progressor can be reset to its original position using the `ResetProgressor` function.

The following Copy Feature Classes to Geodatabase script uses a custom progress dialog box. A step progressor is used, and the number of steps is derived from the number of feature classes in the list. In the `for` loop, the label is changed to the name of the shapefile being copied, and after the shapefile is copied, the step progressor is moved by an increment. The script is as follows:

```
import arcpy
import os
from arcpy import env
env.workspace = arcpy.GetParameterAsText(0)
outworkspace = arcpy.GetParameterAsText(1)
fclist = arcpy.ListFeatureClasses()
fcount = len(fclist)
arcpy.SetProgressor("step", "Copying shapefiles to geodatabase...", 0, ➔
➔ fcount, 1)
for fc in fclist:
    arcpy.SetProgressorLabel("Copying " + fc + "...")
    fcdesc = arcpy.Describe(fc)
    outfc = os.path.join(outworkspace, fcdesc.baseName)
    arcpy.CopyFeatures_management(fc, outfc)
    arcpy.SetProgressorPosition()
arcpy.ResetProgressor()
```

Running the script brings up a progress dialog box with a progressor that shows the percentage completed. This percentage is calculated from the step progressor parameters—that is, the steps are automatically converted to a percentage as they are completed.

Another consideration is the number of steps being used in a step progressor. In many scripts, it is not known in advance how many feature classes, fields, or records will need to be processed. A script that uses a search cursor, for example, may iterate over millions of records. If each iteration were one step, the progress dialog box would need to be updated millions of times, which could severely reduce performance. It may therefore be necessary to include a section in the script that determines the number of iterations (feature classes, rows, or whatever the case may be), and then determines an appropriate number of steps based on the number of iterations.

# 13.12 Running a script in process

Python scripts can be run in process or out of process. Running a script in process means that a script can be run as is without ArcGIS having to start another process or program. Running a script out of process means that ArcGIS has to start another process for the script to run. When another process is started, it takes time for both programs to run, which reduces performance. Other performance issues also arise from message communication between the two processes. In general, therefore, it is recommended that a Python script be run in process so that it will run faster.

Specifying that a tool should run in process or out of process can be done on the Source tab on the tool properties dialog box. By default, this option is selected—that is, scripts are run in process. It should be noted that this option applies only to scripts written in Python.

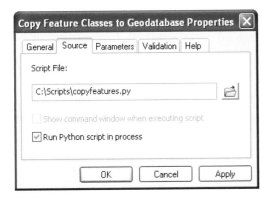

Although running tools in process is recommended to improve performance, there are certain cases when running tools in process can cause problems. For example, some nonstandard modules from third parties may not have the necessary logic to run in process. If you are using third-party modules and are experiencing problems, running the tool out of process may be the solution. Standard Python libraries have modules that have the necessary logic and can be run in process without difficulty.

# Points to remember

- Although Python scripts can be run as stand-alone scripts outside of ArcGIS for Desktop applications, there are many benefits to running scripts as tools. Tools allow a closer integration of scripts in the ArcGIS geoprocessing framework. For example, tools that reference a script can be used in ModelBuilder the same way as any other tool.

- A tool can be created in any custom toolbox and reference a single Python file (.py) that is called when the tool is run.

- For tools to be usable and effective, tool parameters need to be created. This includes setting parameters in the tool properties, as well as including code in the script to receive the parameter values. Tool parameters define what the tool dialog box looks like.

- Effective tools have carefully designed parameters. Each parameter has a data type, such as feature class, table, value, field, or other. The parameter properties provide detailed control of the allowable inputs for each parameter. This ensures that the parameters passed from the tool dialog box to the script are as expected.

- All tools should have outputs so that the tool can be used in ModelBuilder. Sometimes the only way to accomplish this is to use derived parameters, which do not appear on the tool dialog box.

- Tool behavior can be further customized using a `ToolValidator` class.

- Various message functions can be used to write what will appear on the progress dialog box and in the Results window. The appearance of the progress dialog box can be controlled using a number of different functions to change the progressor.

- Running scripts in process is recommended to improve performance.

# Chapter 14
## Sharing tools

## 14.1 Introduction

The ArcGIS geoprocessing framework is designed to facilitate the sharing of tools. Custom toolboxes can be added to ArcToolbox and integrated into regular workflows. Toolboxes can contain any number of tools, consisting of both model tools and script tools. Tools can therefore be shared by distributing a toolbox file (.tbx) that contains the accompanying Python scripts (.py). However, there are a number of obstacles to sharing script tools. One of the principal obstacles is that the resources available to the creator of the script will likely be different from those available to the user of the shared script tools. This includes map documents, toolboxes, scripts, layer files, and any other files used by the tools. Another obstacle is the organization of these resources on a local computer or network. Paths present a fairly persistent problem when sharing tools. This chapter provides guidelines on how to distribute script tools, including how to structure toolboxes, scripts, documentation, and other files that are commonly distributed with shared tools. To help overcome some of these obstacles, ArcGIS 10.1 has introduced geoprocessing packages as a convenient way to distribute shared tools.

## 14.2 Choosing a method for distributing tools

Tools that are developed to share with others can vary from the simple to the complex. The simplest case is a single toolbox file with one or more tools inside the toolbox and no additional files. In a more typical example, a shared tool could consist of a toolbox file, one or more scripts that are used in script tools, and some documentation. A more complex example could contain a toolbox file, several scripts, documentation, compiled Help files, and sample data. A recommended folder structure for these files is presented later in this chapter. A relatively typical folder structure might look like the example in the figure. ➤

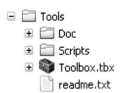

One of the most common ways to share tools is simply to make all the files available in their original folder structure. This typically involves the use of a file compression utility to create a single ZIP file of the folders and their contents. This ZIP file can then be posted online or e-mailed. The recipient can download the file and extract the contents to access the individual folders and files. The toolbox is then added to ArcToolbox to access the tools.

There are two other ways to share tools. If users have access to the same local area network, the folder containing the tools can be copied to a folder that is accessible to all users. A toolbox can be added directly from the network, and no files need to be copied to the user's computer. A second alternative is to publish the toolbox as a geoprocessing service using ArcGIS for Server, which can then be accessed by anyone with an Internet connection.

The method to use depends largely on the relationship between the creator of the tool and the intended users, as well as the software and the skills of the user. For example, if tools are developed primarily for use by others within the same organization, making tools available on a local area network may be the most efficient. To make tools available to a broad community of users, the use of a ZIP file is likely the most convenient.

A number of other considerations will influence how to share tools, including where the input and output data is located and what products and extensions the tools require. In the ZIP file method, for example, any tool data also has to be packaged with the tool because a typical user will not have access to any of the data on the network.

# 14.3 **Handling licensing issues**

Tools distributed using the ZIP file method will run on a user's computer, which may not have the necessary products or licenses to run the tools. Scripts should therefore include logic to check for the necessary product levels (ArcGIS for Desktop Basic, ArcGIS for Desktop Standard, or ArcGIS for Desktop Advanced) and extension licenses (ArcGIS 3D Analyst, ArcGIS Spatial Analyst, and more). Even if a user has the necessary extensions installed, a license may not have been obtained for the current session. In this scenario, the tool will stop with an error message. To facilitate the use of shared tools, the necessary product level and extensions need to be described in the tool's documentation. Working with licenses is covered in chapter 5.

# 14.4 Using a standard folder structure for sharing tools

A standard folder structure, like the example in the figure, is recommended by Esri for easy sharing of geoprocessing tools. There is no requirement to use this specific structure, but it provides a good starting point. →

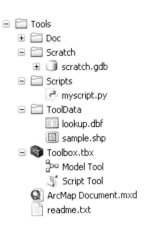

### >>> TIP

Python scripts, by default, are not shown in ArcCatalog, but they can be added as a file type (.py) by going to the menu bar and clicking Customize > ArcCatalog Options > File Types.

The Tools folder contains one or more toolboxes (.tbx files), which contain the tools, including model tools and script tools. Toolboxes can also be placed inside a geodatabase, but a .tbx file directly under the Tools folder is easier to find.

Tools should have the "Store relative path names" check box selected— more on paths later in this chapter. Tool documentation should clearly state what product level and extensions are required for the tools to run. A README file (readme.txt) is often included that explains how the tool works and contains special instructions on how the tool needs to be installed, contact information for the tool's creators, and the like.

Distributing an ArcMap document (.mxd) with the tools is optional but may be helpful if example datasets are part of the shared tool. The ToolData folder may contain sample datasets that a user can work with to learn about the functionality of a tool before trying it out on the user's own data. The tools may also require certain data as part of tool execution, such as lookup tables, also included in this folder.

The Scripts folder contains the Python scripts used in the script tools. Other related files may include script libraries, dynamic-link libraries (DLL), and executable files, such as .exe and .bat (batch) files. Scripts can also be embedded directly into a toolbox, in which case there are no separate script files. This is not very common, however, since in many cases the purpose of sharing the tools is for users to use and learn from the scripts and contribute to their continued improvement.

Many model tools and script tools use a workspace, and a default file geodatabase for scratch data (scratch.gdb) can be provided in the Scratch folder.

Tool documentation is provided in the Doc folder. Documentation can consist of a Microsoft Word document (.doc or .docx) or PDF file (.pdf) that provides instructions, external Help compiled HTML files (.chm) that are referenced by tools or toolboxes, and XML style sheets that replace the default tool dialog boxes and Help dialog boxes. Experienced Python coders are likely to open the actual scripts and learn from both the comments and the code in the scripts. Many other users, however, may never look at the scripts and instead use only the tool dialog boxes. Good documentation will ensure that users get the most out of a tool and understand what it will accomplish, as well as its limitations, without having to open the actual scripts.

# 14.5 **Working with paths**

Paths are an integral part of working with data and tools. When tools are shared, paths become particularly important, because without proper documentation of where files are located, the tools will not run.

If you have worked with ArcGIS to create map documents or tools, you are probably familiar with absolute and relative paths. Absolute paths are also referred to as full paths. They start with a drive letter, followed by a colon (:), and then the folder and file name—for example, C:\Data\streams.shp. Relative paths refer to a location that is relative to a current folder. Relative paths make use of special symbols—a single dot (.) and a double dot (..). A single dot represents the folder you are working in, and a double dot represents the parent folder. Although technically correct, this convention for navigating folders is not very practical because you cannot type relative paths in ArcGIS or Python scripts. Still, it is important to understand the concept of relative paths and what it means in respect to manipulating data in ArcGIS.

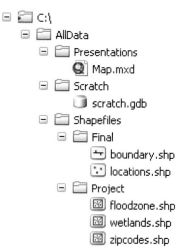

Consider the following example with two shapefiles located in the C:\alldata\shapefiles\final folder: boundary.shp and locations.shp. Relative to each other, there is no need to know the path other than the base names—that is, the file names. Now consider an example where you want to run a tool that uses the shapefiles locations.shp and floodzone.shp. These files are located in two different folders, and therefore their relative paths are final\locations.shp and project\floodzone.shp. The higher-level folders—that is, alldata\shapefiles–are not needed to locate one file relative to the other. ➜

You have likely worked with relative paths when saving map documents (.mxd files). To avoid lost data connections when folders are moved or renamed, the data source options in map document properties can be set to relative paths.

| Pathnames: | ☑ Store relative pathnames to data sources |
|---|---|

When map documents are saved using relative paths, ArcMap converts absolute paths to relative paths based on the location where the .mxd file is stored. For example, consider a map document that is saved with relative paths enabled and that is stored as follows:

```
C:\alldata\presentations\Map.mxd
```

If the shapefile locations.shp is added to this map document, the absolute path is converted to the following relative path:

```
.\..\shapefiles\final\locations.shp
```

Relative to the location of the map document (Map.mxd), locations.shp is located in the parent folder of the map document—that is, C:\alldata (hence, the single dot followed by the double dot)—and under the subfolder shape-files\final. The name and location of the parent folder itself is not needed to access the shapefile and is therefore not part of the relative path.

### >>> TIP

Don't worry too much about the notation for relative paths since you can't type this notation in ArcGIS or Python anyway.

The use of relative paths makes it possible to move or rename folders. For example, if the alldata folder were renamed "data," all paths in the ArcMap document would remain intact. Similarly, if the drive letter were modified from C to E, all paths would also remain intact.

One limitation of relative paths is that they cannot span multiple disk drives. If some files are located on the C drive and some on the E drive, only absolute paths will preserve the correct locations of all files.

Similar to working with map documents, absolute paths and relative paths can be used in model tools. Relative paths for models can be enabled on the model properties dialog box. ➔

In script tools, relative paths are enabled in the Add Script wizard when creating a script tool, or on the script tool properties dialog box for existing tools. ➔

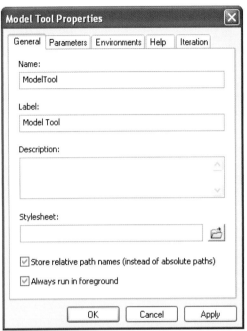

Relative paths are relative to the current folder where the toolbox file is located. When relative paths are enabled, it applies to the script files, datasets used for the default value properties, files referenced in the tool documentation, layer files used for the symbology properties, compiled Help files, and style sheets.

It is important to recognize that paths within the script are not converted because ArcGIS has no reliable way to examine and modify the script code. Therefore, if a script uses absolute paths, they are not converted to relative paths when relative paths are enabled for the script tool.

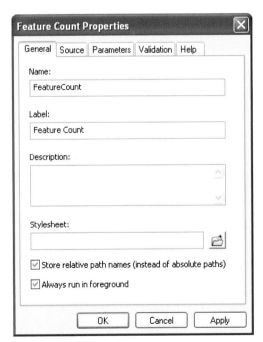

### >>> TIP

In general, Python code needs to be written so that files can be found relative to a known location, which is typically the location of the script itself.

After this review of working with paths, it is worthwhile to revisit relative paths in the context of sharing tools. For the purpose of this discussion, the same example folder structure discussed earlier in this chapter, as shown in the figure, will be used. ➜

To share tools, relative paths have to be enabled in the script tool properties. In this example, the script tool will reference a script in the Scripts folder. It may also reference tool documentation in the Doc folder. The script itself may reference data in the ToolData folder. These references will continue to be valid when the script tool is shared with another user, as long as the standard folder structure is maintained. If the toolbox file (Toolbox.tbx) containing the script tool were moved to a different location separate from the other folders and files, the script files called by the script tool would not be found and the script would not work. The tool dialog box would open, but upon tool execution, the following error message would appear: "Script associated with this tool does not exist." Therefore, for a script tool to work correctly, the folder structure must be maintained.

```
☐ 📁 Tools
   ⊞ 📁 Doc
   ☐ 📁 Scratch
      ⊞ 🛢 scratch.gdb
   ☐ 📁 Scripts
         myscript.py
   ☐ 📁 ToolData
         lookup.dbf
         sample.shp
   ☐ 🧰 Toolbox.tbx
         Model Tool
         Script Tool
      ArcMap Document.mxd
      readme.txt
```

## 14.6 Finding data and workspaces

In general, it is best to avoid hard-coding paths into your script if it is going to be shared with others as a script tool. Instead, the paths are derived from the parameters on the tool dialog box and these paths are passed to the script. The script reads these parameters using GetParameterAsText and related functions.

Sometimes, however, it is necessary to hard-code paths to a particular location. For example, an existing layer file may be necessary to set the symbology for an output parameter. Or a particular tool may require the use of a lookup table. Depending on the nature of the information, it may already be incorporated into the script (for example, a lookup table can be coded as a Python dictionary), but this may not always be possible. Some data files may therefore be necessary for the tool to run, even though they are not provided as parameters for the user. Instead, these data files are provided by the author of the script and distributed as part of the shared tool. Following the suggested folder structure presented earlier in this chapter, these files would be placed in the ToolData folder, making it possible for the data files to be found relative to the location of the script.

The path of the script can be found using the following code:

```
scriptpath = sys.path[0]
```

Or:

```
scriptpath = os.getcwd()
```

Running this code results in a string with the complete path of the script, but without the name of the script itself. If the data files necessary for the script to run are located in the ToolData folder, per the suggested folder structure, the Python module os.path can be used to create a path to the data.

The folder structure used thus far can serve as an example. The Tools folder contains the shared tools, including the toolbox, the script, and the data files. Relative paths are enabled for the script tool, so the Tools folder can be moved, or even renamed, and the script will still work. The script referenced by the script tool is located in the Scripts folder. The script needs a file called lookup.dbf, located in the ToolData folder, to run. The file name can be hard-coded into the script because the author of the script is also the author of the lookup.dbf file and the creator of the ToolData folder. However, the absolute path should not be hard-coded into the script, but the relative path used instead: ToolData\lookup.dbf. This will make it possible for the Tools folder to be moved to any location without the user of the script tool being limited to the absolute path originally used by the author of the script.

The code that references the lookup.dbf file in the script is as follows:

```
import arcpy
import os
import sys
scriptpath = sys.path[0]
toolpath = os.path.dirname(scriptpath)
tooldatapath = os.path.join(toolpath, "ToolData")
datapath = os.path.join(tooldatapath, "lookup.dbf")
```

Notice that two elements are hard-coded into the script: the actual file name of the tool data (lookup.dbf) and the folder where the tool data is located (ToolData). These are both created by the author of the tool and can therefore be hard-coded into the script because they do not depend on user input.

A similar approach can be used to reference the location of a scratch workspace. Scratch workspaces are very common in models, and they can also be used in scripts. Using the same example folder structure as before, the script to set the scratch workspace is as follows:

```
import arcpy
import os
import sys
from arcpy import env
scratchws = env.scratchWorkspace
scriptpath = sys.path[0]
toolpath = os.path.dirname(scriptpath)
if not env.scratchWorkspace:
    scratchws = os.path.join(toolpath, "Scratch/scratch.gdb")
```

When a scratch workspace is set, a few other considerations should be kept in mind, as follows:

- Write permission to the workspace is required.

- A scratch workspace can be set as part of the environment settings. If this is the case, a script should use this workspace because it most likely has been intentionally set by the user. So the preceding example code would typically be preceded by an `if` statement that evaluates whether a scratch workspace has been specified in the environment settings that are being passed to the script.

- Using the current workspace as a scratch workspace is possible, but it can cause problems. First, the current workspace can become cluttered if the script generates a lot of output. Second, cleaning up the results afterward can be cumbersome because it may be difficult to separate intermediate data, which can be deleted, from the final results. Third, if the current workspace is a remote database on a server, it can cause performance issues.

Finally, instead of the name of an (empty) scratch geodatabase being hard-coded into a script, the `CreateScratchName` function can be used to create a unique dataset in the scratch workspace.

## 14.7 Creating a geoprocessing package

The approach for distributing shared tools as described so far is quite robust but also somewhat cumbersome. It typically requires that you manually consolidate data, tools, and supporting files into a single folder. As an alternative, ArcGIS 10.1 has introduced geoprocessing packages, which are a more convenient way to distribute all the tools and files related to geoprocessing workflows. This section describes what a geoprocessing package is and how to create it.

A geoprocessing package is a single compressed file with a .gpk extension. This single file contains all the files necessary to run a particular geoprocessing workflow, including custom tools, input datasets, and other supporting files. This file can be posted online, e-mailed, or shared through a local area network. Although this sounds similar to the use of a ZIP file, as described earlier in this chapter, geoprocessing packages are created very differently and have additional functionality.

A geoprocessing package is created from one or more results in the Results window, which have been created by successfully running one or more geoprocessing tools. A basic workflow to create and share a geoprocessing package is as follows:

1. Add data (and custom tools if needed) to an ArcGIS for Desktop application.

2. Create a geoprocessing workflow by running one or more tools.

3. In the Results window, select one or more results, right-click the selection, and click Share As > Geoprocessing Package.

4. Complete the entries on the Geoprocessing Package dialog box, which includes options for sharing, for adding additional results or files, and for sharing datasets or only the schema.

5. Share the resulting .gpk file.

A recipient of the geoprocessing package can open the contents in an ArcGIS for Desktop application to examine the datasets and workflows used. A single .gpk file contains all the resources needed to rerun the geoprocessing workflow, including tools, layers, and other files. Tools can include system tools as well as custom tools. So, if a geoprocessing result was created using a script tool, this script tool and the underlying .py files necessary for the tool to run would all be included in the geoprocessing package.

The single greatest benefit of using geoprocessing packages is that all the necessary resources are automatically combined in a single file, no matter where they are located. There is no need to manually consolidate all the resources into a single folder as required by the more traditional approach using ZIP files.

Geoprocessing packages are described in great detail in ArcGIS Desktop Help on the Contents tab, under Geoprocessing > Sharing geoprocessing workflows.

## 14.8 Embedding scripts and password-protecting tools

The most common way to share script tools is to reference the Python script file in the script tool properties and to provide the script file separately, typically in the Scripts folder. This allows users to clearly see which scripts are being used, and the scripts can be opened to view the code.

Scripts can also be embedded in a toolbox. The code is then contained within the toolbox, and a separate script file is no longer needed. This approach can make it easier to manage and share tools.

To import a script, right-click the script tool and click Import Script. Once a script is imported into a tool, the toolbox can be shared without including the script files. In other words, just sharing the .tbx file is sufficient, and no separate .py files need to be provided for the script tool to run.

When a script is imported, however, the original script file is not deleted—it is simply copied and embedded in the toolbox.

Embedding scripts does not mean they can no longer be viewed or edited. Say, for example, you have imported a script and shared a toolbox with another user. The recipient can right-click the script tool and click Export to obtain a copy of the original script. Once exported, the script can be viewed and edited the same way as any other script in a shared tool. Although embedding scripts is a useful way to reduce the number of files to manage and share, it can lead to some confusion. For example, some script tools use multiple scripts—for example, a script that is referenced by the script tool and additional scripts that are called by the first script. Embedding multiple scripts can be confusing to users because it becomes less transparent how the scripts work. In addition, embedding scripts was introduced in ArcGIS 10, so many users are probably more familiar with seeing a .tbx file and one or more .py files than having them embedded.

One very good reason to embed your scripts is to create password protection. Regular script files cannot be password protected. If you share your tools including individual .py files, any user can open these scripts with a Python editor or a text editor. Users can modify the code or copy it for use in their own scripts. This is, in fact, one of the reasons why working with Python is so appealing. If, for some reason, you need password protection, you can right-click the script tool and click Set Password—this works only if you have previously imported a script. →

Setting a password does not affect execution of the script tool, but any attempt to view the script or export the script will prompt the use of a password.

# 14.9 **Documenting tools**

Good documentation is important when sharing tools. Documentation includes background information on how the tool was developed as well as specifics on how the tool works. Documentation can also be used to explain specific concepts that may be new to other users.

There are a number of ways to provide documentation for a tool within ArcGIS, as follows:

- A brief, text-only description can be provided on the tool properties dialog box. This description will appear as the default text in the Help panel on the tool dialog box.

- A more detailed description can be created by editing the Description page of a tool. This description is used in multiple locations, including the tool reference page, the Help panel on the tool dialog box, and the Help panel in the Python window.

- A style sheet can be used, providing additional control of the look and feel of the tool dialog box.

- A compiled Help file can be created and referenced to appear as the tool reference page.

There are other ways to provide documentation as well, within the script itself or on disk, as follows:

- By commenting code. Good scripts contain detailed comments, which explain how a script works. Not all users of a script tool may look at the script, but for those who do, comments can be very informative. Comments are located inside the actual script files.

- Through separate documentation located on disk—for example, in the Doc folder. Documentation files can be provided as .doc, .docx, .pdf, or other file types. This documentation typically includes a more detailed explanation of the tools and any relevant background concepts.

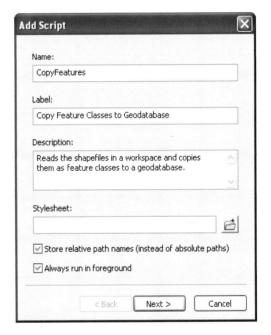

Following is a more detailed explanation of some of the ways to provide documentation for a tool.

## Providing a brief description

The simplest form of tool documentation in ArcGIS is accomplished by providing a brief description in the Description box on the tool properties dialog box. This description can be created in the Add Script wizard or by accessing the script tool properties. ➔

By default, this description will appear in the Help panel on the tool dialog box.

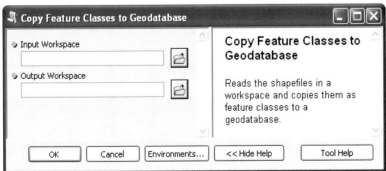

This type of tool documentation is a bit limited because it allows for text only, and it provides a description of the tool only, and not individual parameters. However, it is a good place to start.

## Editing the Description page

By default, when a script tool is created, a Description page is also created, which is populated with the tool's basic syntax. This page can be viewed in ArcCatalog on the Description tab, where you would normally review metadata. For example, the default Description page of the Copy Feature Classes to Geodatabase script tool created in chapter 13 is shown in the figure. ➜

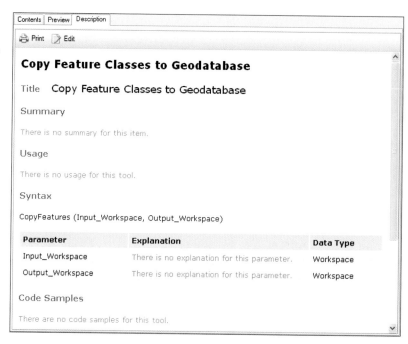

The default description is rather rudimentary and provides little in addition to what the tool dialog box does. However, the contents of the Description page can be modified in ArcCatalog by clicking the Edit button under the Description tab, in the same way that metadata can be edited. ➜

From here, you can edit the title and summary, provide examples of the usage, enter details of the syntax in a table, and add code samples. You can also load a thumbnail graphic that illustrates how the tool works. Tags are used to identify the subject or content of your tool. The documentation you provide for Item Description is used by the Search window to find your tool.

Notice that there are many entries in the description that are not relevant to tool usage because they are mostly intended for documenting the metadata for datasets.

Once you modify the Description page, this information is used on the tool dialog box. For example, if you enter a description for individual parameters, it will appear on the tool dialog box Help panel for that parameter. ➜

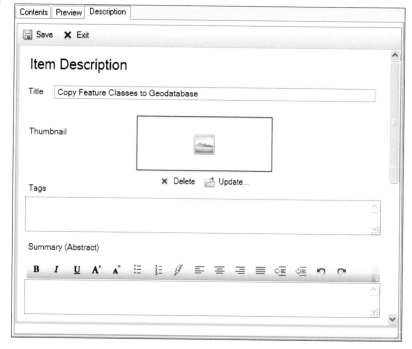

The Description page also becomes the default page that loads when the Help option for a particular tool is clicked, unless a different file has been specified.

**Using style sheets**

Style sheets are used to control the properties of the tool dialog box. A style sheet provides style and layout information, including fonts, alignment, and margins. The default style sheet is typically sufficient and is found in the C:\Program Files\ArcGIS\Desktop10.1\ArcToolbox\Stylesheets folder. The default style sheet is applied automatically for any new tool dialog box, but you can create your own if desired.

*Note: Creating your own style sheet is not covered in this book.*

**Using a compiled Help file**

A script tool can reference a compiled Help file (.chm). This file is used when viewing the tool Help page. Compiled Help files are similar to HTML files but are compiled to create a single, self-contained package of documentation. Compiled Help files are a proprietary format for Help files developed by Microsoft for use in the Windows operating system. Creating a compiled Help file requires the use of the Microsoft Help Software Development Kit (SDK).

*Note: Creating a compiled Help file is not covered in this book.*

If you have a compiled Help file, you can reference it on the Help tab on the tool properties dialog box. Optionally, you can provide the Help context by specifying an HTML topic ID. The Help context controls which topic in the .chm file will be displayed. ➜

If you have already created detailed documentation on the tool's Description page, you can export this documentation to an HTML file. This is an option on the Help tab on the tool properties dialog box. You can then use this HTML file when creating a compiled Help file.

# 14.10 **Example tool: Market analysis**

This section looks at an example tool to review the organization of the files that are part of the tool, as well as the tool's documentation. The example tool is a market analysis tool based on the Huff Model, which is used to estimate sales potential for store locations based on demographic data. The tool was created by Drew Flater as an example of a more advanced script tool and is posted on the ArcGIS for Desktop Resource Center.

The tool's organization closely follows the suggested folder structure. There is a single .tbx file with a single script tool. The actual script file, HuffModel.py, is located in the Script folder. The ToolData folder contains a sample dataset as well as some data that the tool needs for execution. The Doc folder contains a .doc or .docx file and a .pdf file that describe the sample data and includes a brief tutorial on using the tool. ➔

Detailed tool documentation can be found on the Description tab in ArcCatalog. The same documentation can be accessed by viewing the tool's Help file, although the formatting will be slightly different depending on the HTML browser. The text and formatting of this documentation is stored in the toolbox. The image in the documentation is referenced in the tool's description, but the actual image file, help.png, is located in the Doc folder.

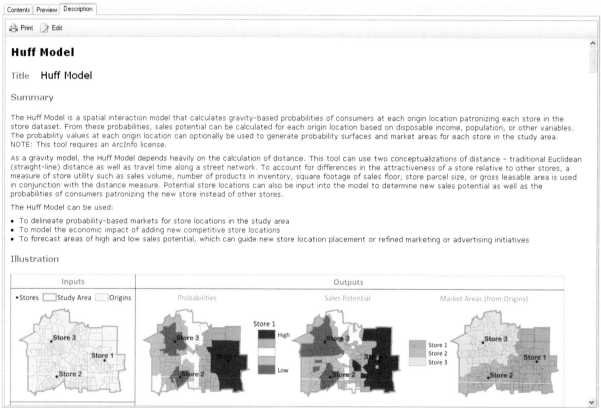

The tool dialog box itself contains a large number of required and optional parameters. When the tool dialog box is open, a brief description of the tool appears in the Help panel.

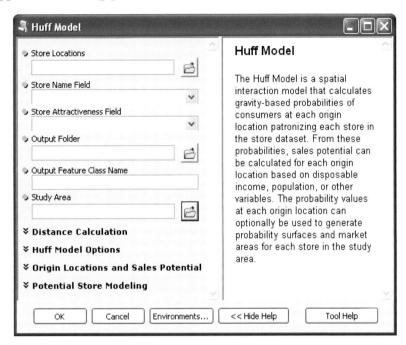

The content of the Help panel changes when the cursor is placed inside one of the parameter boxes—in this case, the Store Attractiveness field. This information is derived from the tool's description.

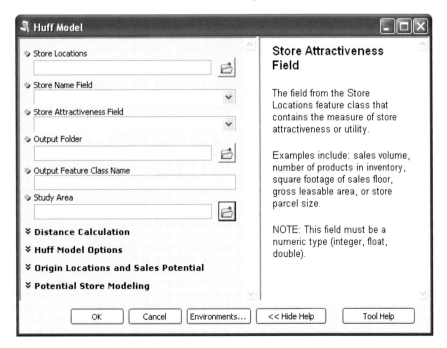

Separate from ArcGIS, tool documentation is provided in the form of a tutorial file on disk that describes the sample data and explains how to run the tool.

# Market Analysis with the Huff Model tool

## Sample Data

Sample data stored at ...MarketAnalysisToolbox\ToolData\SAMPLE\Sample.gdb

1. Feature Class 'Stores' contains three point features representing retail stores arbitrarily located in the study area for demonstration purposes – they do not represent real store locations. 'Stores' will be used in the "Store Locations" parameter of the Huff Model tool. 'Stores' contains fields "Name" and "Sales" which will be used in the "Store Name Field" and "Store Attractiveness Field" parameters of the Huff Model tool, respectively.

2. Feature Class 'Study_Area' contains a single polygon feature roughly centered on the urban area of Akron, Ohio, United States. 'Study_Area' will be used in the "Study Area" parameter of the Huff Model tool.

3. Feature Class 'Block_Groups' contains 189 polygon features which are U.S. Census Bureau block groups covering the same urban area of Akron, Ohio, United States. 'Block_Groups' can optionally be used in the "Origin Locations" parameter of the Huff Model tool (under the Origin Locations and Sales Potential category). 'Block_Groups' has a number of demographic indicator fields, one of which can optionally be used in the "Sales Potential Field" parameter of the Huff Model tool (under the Origin Locations and Sales Potential category). Suggested fields are "POP2007" or "HOUSEHOLDS".

## Market Analysis Tutorial

**1.** Add the above feature classes to a new ArcMap document. Add MarketAnalysisTools.tbx to the ArcToolbox window. Open the Huff Model tool from the Market Analysis Tools toolbox. Opening each of the drop-down categories, the tool dialog should appear as below.

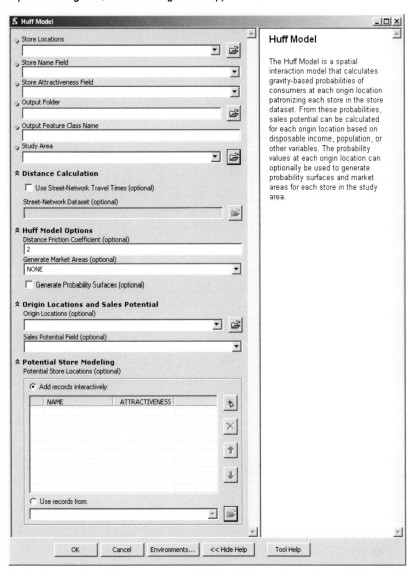

Finally, the script itself contains documentation in the form of comments.

```
# -------------------------------------------------------------------
# HuffModel.py
# Created: 4/13/2007 by Drew Flater
# Usage: Creating probability-based trade areas for retail stores
# -------------------------------------------------------------------

# Import system modules
import sys, string, arcgisscripting, os, traceback, shutil, re

# Create the Geoprocessor object
gp = arcgisscripting.create(93)

# Set overwrite
gp.overwriteoutput = 1

def AddPrintMessage(msg, severity):
    print msg
    if severity == 0: gp.AddMessage(msg)
    elif severity == 1: gp.AddWarning(msg)
    elif severity == 2: gp.AddError(msg)

# Start traceback Try-Except statement:
try:
    # Script parameters...
    stores = gp.getparameterastext(0)
    store_name = gp.getparameterastext(1)
    store_attr = gp.getparameterastext(2)
    outfolder = gp.getparameterastext(3)
    fc_name = gp.getparameterastext(4)
    studyarea = gp.getparameterastext(5)
    blockgroups = gp.getparameterastext(6)
```

This example script is relatively advanced—it has 748 lines of code, and the tool dialog box has 18 parameters. But the tool's standard organization and documentation make it relatively easy to use.

# Points to remember

- The ArcGIS geoprocessing framework is designed to facilitate the sharing of tools. Custom toolboxes can be added to ArcToolbox and integrated into regular workflows. Toolboxes can contain any number of tools, consisting of both model tools and script tools. Tools can therefore be shared by distributing a toolbox file (.tbx) that contains the accompanying Python scripts (.py) and any other resources needed to run the tools.

- To ensure custom tools work properly, the resources needed to run the tools should be made available in a standard folder structure. This includes folders for scripts, data, and documentation.

- Absolute paths work only when files are not moved and folders are not renamed. To share tools, relative paths should be enabled for each script tool. Relative paths are relative to the current folder, which for scripts is where the toolbox is located. Relative paths cannot span multiple drives.

- Geoprocessing packages provide an alternative way to distribute script tools. A geoprocessing package is a single, compressed file with a .gpk extension that contains all the files necessary to run a particular geoprocessing workflow, including custom tools, input datasets, and other supporting files.

- Shared tools can be documented in various ways, including editing the Description page in ArcCatalog, using style sheets, and referencing compiled Help files.

# Appendix A
## Data source credits

## Exercise 1
None

## Exercise 2
\EsriPress\Python\Data\Exercise02\basin.shp, courtesy of Clay County, Minnesota
\EsriPress\Python\Data\Exercise02\floodzones.shp, courtesy of Clay County, Minnesota
\EsriPress\Python\Data\Exercise02\lakes.shp, courtesy of Clay County, Minnesota
\EsriPress\Python\Data\Exercise02\rivers.shp, courtesy of Clay County, Minnesota
\EsriPress\Python\Data\Exercise02\roads.shp, courtesy of Clay County, Minnesota
\EsriPress\Python\Data\Exercise02\soils.shp, courtesy of Clay County, Minnesota

## Exercise 3
\EsriPress\Python\Data\Exercise03\zipcodes.shp, courtesy of City of Austin, Texas

## Exercise 4
None

## Exercise 5
\EsriPress\Python\Data\Exercise05\bike_routes.shp, courtesy of City of Austin, Texas
\EsriPress\Python\Data\Exercise05\facilities.shp, courtesy of City of Austin, Texas
\EsriPress\Python\Data\Exercise05\hospitals.shp, courtesy of City of Austin, Texas
\EsriPress\Python\Data\Exercise05\parks.shp, courtesy of City of Austin, Texas
\EsriPress\Python\Data\Exercise05\zip.shp, courtesy of City of Austin, Texas

## Exercise 6

\EsriPress\Python\Data\Exercise06\amtrak_stations.shp,
   courtesy of The National Atlas of the United States of America
\EsriPress\Python\Data\Exercise06\cities.shp, courtesy of The National Atlas of the United States of America
\EsriPress\Python\Data\Exercise06\counties.shp, courtesy of The National Atlas of the United States of America
\EsriPress\Python\Data\Exercise06\new_mexico.shp,
   courtesy of The National Atlas of the United States of America
\EsriPress\Python\Data\Exercise06\railroads.shp, courtesy of The National Atlas of the United States of America

## Exercise 7

\EsriPress\Python\Data\Exercise07\airports.shp, courtesy of The National Atlas of the United States of America
\EsriPress\Python\Data\Exercise07\alaska.shp, courtesy of The National Atlas of the United States of America
\EsriPress\Python\Data\Exercise07\roads.shp, courtesy of The National Atlas of the United States of America

## Exercise 8

\EsriPress\Python\Data\Exercise08\coordinates.txt, created by the author
\EsriPress\Python\Data\Exercise08\dams.shp, courtesy of The National Atlas of the United States of America
\EsriPress\Python\Data\Exercise08\hawaii.shp, courtesy of The National Atlas of the United States of America
\EsriPress\Python\Data\Exercise08\rivers.shp, courtesy of The National Atlas of the United States of America

## Exercise 9

\EsriPress\Python\Data\Exercise09\elevation, courtesy of US Geological Survey
\EsriPress\Python\Data\Exercise09\landcover.tif, courtesy of US Geological Survey
\EsriPress\Python\Data\Exercise09\tm.img, courtesy of US Geological Survey

## Exercise 10

\EsriPress\Python\Data\Exercise10\counties.shp, from Esri Data & Maps 2009,
   from Esri, derived from Tele Atlas, US Census, Esri (Pop2008 field)
\EsriPress\Python\Data\Exercise10\Austin\addresses.shp, courtesy of City of Austin, Texas
\EsriPress\Python\Data\Exercise10\Austin\base.shp, courtesy of City of Austin, Texas
\EsriPress\Python\Data\Exercise10\Austin\buildings.shp, courtesy of City of Austin, Texas
\EsriPress\Python\Data\Exercise10\Austin\facilities.shp, courtesy of City of Austin, Texas
\EsriPress\Python\Data\Exercise10\Austin\historical_landmarks.shp, courtesy of City of Austin, Texas
\EsriPress\Python\Data\Exercise10\Austin\hospitals.shp, courtesy of City of Austin, Texas
\EsriPress\Python\Data\Exercise10\Austin\parks.shp, courtesy of City of Austin, Texas
\EsriPress\Python\Data\Exercise10\Austin\sidewalks.shp, courtesy of City of Austin, Texas
\EsriPress\Python\Data\Exercise10\Austin\trees.shp, courtesy of City of Austin, Texas

## Exercise 11

\EsriPress\Python\Data\Exercise11\bike_routes.shp, courtesy of City of Austin, Texas
\EsriPress\Python\Data\Exercise11\county.shp, courtesy of City of Austin, Texas
\EsriPress\Python\Data\Exercise11\facilities.shp, courtesy of City of Austin, Texas
\EsriPress\Python\Data\Exercise11\parks.shp, courtesy of City of Austin, Texas

## Exercise 12

\EsriPress\Python\Data\Exercise12\parcels.shp, created by the author
\EsriPress\Python\Data\Exercise12\streets.shp, courtesy of Clay County, Minnesota

## Exercise 13

\EsriPress\Python\Data\Exercise13\points.shp, courtesy of City of Austin, Texas

## Exercise 14

None

# Appendix B
## Data license agreement

*Important:* *Read carefully before opening the sealed media package.*

Environmental Systems Research Institute, Inc. (Esri), is willing to license the enclosed data and related materials to you only upon the condition that you accept all of the terms and conditions contained in this license agreement. Please read the terms and conditions carefully before opening the sealed media package. By opening the sealed media package, you are indicating your acceptance of the Esri License Agreement. If you do not agree to the terms and conditions as stated, then Esri is unwilling to license the data and related materials to you. In such event, you should return the media package with the seal unbroken and all other components to Esri.

### Esri License Agreement
This is a license agreement, and not an agreement for sale, between you (Licensee) and Environmental Systems Research Institute, Inc. (Esri). This Esri License Agreement (Agreement) gives Licensee certain limited rights to use the data and related materials (Data and Related Materials). All rights not specifically granted in this Agreement are reserved to Esri and its Licensors.

### Reservation of Ownership and Grant of License
Esri and its Licensors retain exclusive rights, title, and ownership to the copy of the Data and Related Materials licensed under this Agreement and, hereby, grant to Licensee a personal, nonexclusive, nontransferable, royalty-free, worldwide license to use the Data and Related Materials based on the terms and conditions of this Agreement. Licensee agrees to use reasonable effort to protect the Data and Related Materials from unauthorized use, reproduction, distribution, or publication.

### Proprietary Rights and Copyright
Licensee acknowledges that the Data and Related Materials are proprietary and confidential property of Esri and its Licensors and are protected by United States copyright laws and applicable international copyright treaties and/or conventions.

## Permitted Uses

Licensee may install the Data and Related Materials onto permanent storage device(s) for Licensee's own internal use.

Licensee may make only one (1) copy of the original Data and Related Materials for archival purposes during the term of this Agreement unless the right to make additional copies is granted to Licensee in writing by Esri.

Licensee may internally use the Data and Related Materials provided by Esri for the stated purpose of GIS training and education.

## Uses Not Permitted

Licensee shall not sell, rent, lease, sublicense, lend, assign, time-share, or transfer, in whole or in part, or provide unlicensed Third Parties access to the Data and Related Materials or portions of the Data and Related Materials, any updates, or Licensee's rights under this Agreement.

Licensee shall not remove or obscure any copyright or trademark notices of Esri or its Licensors.

## Term and Termination

The license granted to Licensee by this Agreement shall commence upon the acceptance of this Agreement and shall continue until such time that Licensee elects in writing to discontinue use of the Data or Related Materials and terminates this Agreement. The Agreement shall automatically terminate without notice if Licensee fails to comply with any provision of this Agreement. Licensee shall then return to Esri the Data and Related Materials. The parties hereby agree that all provisions that operate to protect the rights of Esri and its Licensors shall remain in force should breach occur.

## Disclaimer of Warranty

The Data and Related Materials contained herein are provided "as-is," without warranty of any kind, either express or implied, including, but not limited to, the implied warranties of merchantability, fitness for a particular purpose, or noninfringement. Esri does not warrant that the Data and Related Materials will meet Licensee's needs or expectations, that the use of the Data and Related Materials will be uninterrupted, or that all nonconformities, defects, or errors can or will be corrected. Esri is not inviting reliance on the Data or Related Materials for commercial planning or analysis purposes, and Licensee should always check actual data.

## Data Disclaimer

The Data used herein has been derived from actual spatial or tabular information. In some cases, Esri has manipulated and applied certain assumptions, analyses, and opinions to the Data solely for educational training purposes. Assumptions, analyses, opinions applied, and actual outcomes may vary. Again, Esri is not inviting reliance on this Data, and the Licensee should always verify actual Data and exercise their own professional judgment when interpreting any outcomes.

## Limitation of Liability

Esri shall not be liable for direct, indirect, special, incidental, or consequential damages related to Licensee's use of the Data and Related Materials, even if Esri is advised of the possibility of such damage.

## No Implied Waivers

No failure or delay by Esri or its Licensors in enforcing any right or remedy under this Agreement shall be construed as a waiver of any future or other exercise of such right or remedy by Esri or its Licensors.

## Order for Precedence

Any conflict between the terms of this Agreement and any FAR, DFAR, purchase order, or other terms shall be resolved in favor of the terms expressed in this Agreement, subject to the government's minimum rights unless agreed otherwise.

## Export Regulation

Licensee acknowledges that this Agreement and the performance thereof are subject to compliance with any and all applicable United States laws, regulations, or orders relating to the export of data thereto. Licensee agrees to comply with all laws, regulations, and orders of the United States in regard to any export of such technical data.

## Severability

If any provision(s) of this Agreement shall be held to be invalid, illegal, or unenforceable by a court or other tribunal of competent jurisdiction, the validity, legality, and enforceability of the remaining provisions shall not in any way be affected or impaired thereby.

## Governing Law

This Agreement, entered into in the County of San Bernardino, shall be construed and enforced in accordance with and be governed by the laws of the United States of America and the State of California without reference to conflict of laws principles. The parties hereby consent to the personal jurisdiction of the courts of this county and waive their rights to change venue.

## Entire Agreement

The parties agree that this Agreement constitutes the sole and entire agreement of the parties as to the matter set forth herein and supersedes any previous agreements, understandings, and arrangements between the parties relating hereto.

# Appendix C
## Installing the data and software

Python Scripting for ArcGIS includes a DVD containing data and exercises. A free, fully functioning 180-day trial version of ArcGIS 10.1 for Desktop Advanced software can be downloaded at **esri.com/pythonscriptingforArc-GIS10-1**. You will find an authorization number printed on the inside back cover of this book. You will use this number when you are ready to install the software.

If you already have a licensed copy of ArcGIS 10.1 for Desktop Advanced software installed on your computer (or have access to the software through a network), do not install the trial software. Use your licensed software to do the exercises in this book. If you have an older version of ArcGIS software installed on your computer, you must uninstall it before you can install the software that is provided with this book.

.NET Framework 3.5 SP1 must be installed on your computer before you install ArcGIS 10.1 for Desktop software. Some features of ArcGIS 10.1 for Desktop software require Windows Internet Explorer version 8.0. If you do not have Internet Explorer version 8.0, you must install it before installing the ArcGIS 10.1 for Desktop software.

## Installing the exercise data

Follow the steps below to install the exercise data.

1. Put the data DVD into your computer's DVD drive. A splash screen will appear.

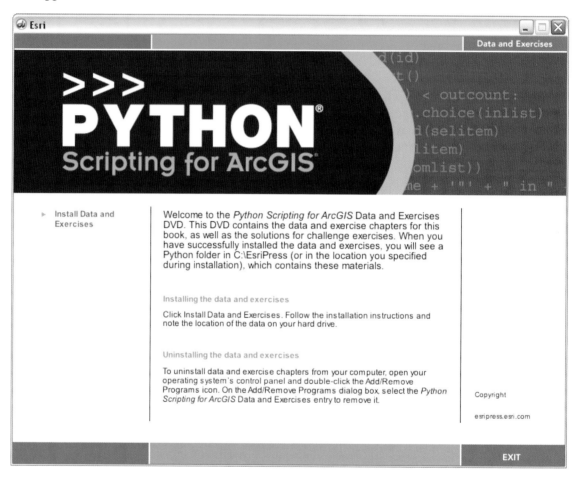

2. Read the welcome, and then click the Install Resources link. This launches the InstallShield Wizard.

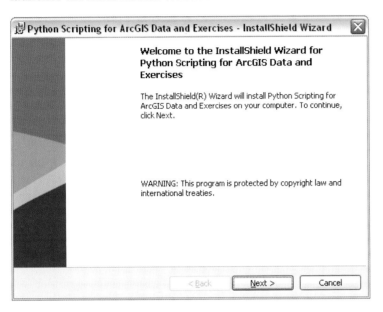

3. Click Next. Read and accept the license agreement terms, and then click Next.

4. Accept the default installation folder or click Browse and navigate to the drive or folder location where you want to install the data.

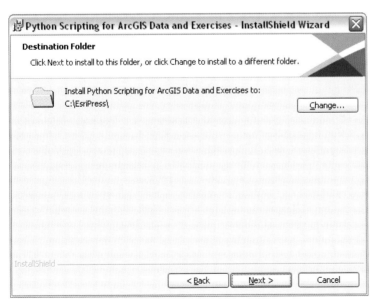

5.  Click Next. The installation will take a few moments. When the installation is complete, you will see the following message.

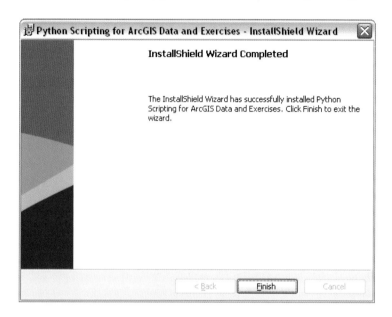

6.  Click Finish. The exercise data is installed on your computer in a folder called C:\EsriPress\Python\Data.

## Uninstalling the data and resources

To uninstall the data and resources from your computer, open your operating system's control panel and double-click the Add/Remove Programs icon. In the Add/Remove Programs dialog box, select the following entry and follow the prompts to remove it:

Python Scripting for ArcGIS Data and Exercises

# Installing the software

*Note: If you already have a licensed copy of ArcGIS 10.1 for Desktop Advanced software installed on your computer or have access to the software through a network, use it to do the exercises in this book. If you need a copy of ArcGIS 10.1 for Desktop Advanced software, you can obtain a free trial version from Esri. This trial version is intended for educational purposes only and will expire 180 days after you install and register the software. The software cannot be reinstalled nor can the time limit be extended. It is recommended that you uninstall this software when it expires.*

Follow these steps to obtain the 180-day free trial of ArcGIS 10.1 for Desktop Advanced software:

1. Check the system requirements for ArcGIS to make sure your computer has the hardware and software required for the trial: esri.com/arcgis101sysreq.

2. Uninstall any previous versions of ArcGIS Desktop that you already have on your computer. For best results, please download and use the ArcGIS Uninstall Utility from the download page.

3. Go to esri.com/pythonscriptingforArcGIS10-1, and then follow the instructions for obtaining the software.

*Note: You must have an Esri Global Account to receive your free trial software. Click Create an account if you do not have one. When prompted, enter your 12-character authorization number (EVAxxxxxxxxx) printed on the inside back cover of the book.*

Assistance, FAQs, and support for your trial software are available on the online resources page at esri.com/trialhelp.

# Uninstalling the software

To uninstall the software from your computer, use Add/Remove Programs from your operating system's control panel. Select the following entry and follow the prompts to remove it:

ArcGIS 10.1 for Desktop

# Index